W9-CJM-616

Cosmic Magnetism

Cosmic Magnetism

Percy Seymour
Principal Lecturer in Astronomy
Plymouth Polytechnic

Adam Hilger

Adam Hilger, Bristol and Boston

British Library Cataloguing in Publication Data

Seymour, Percy
Cosmic magnetism.
1. Magnetic fields (Cosmic physics)
I. Title
523.1′1 QC809.M25

ISBN 0-85274-556-7

Published under the Adam Hilger imprint by IOP Publishing Ltd
Techno House, Redcliffe Way, Bristol BS1 6NX, England
PO Box 230, Accord, MA 02018, USA

Typeset by BC Typesetting, Bristol BS15 6PJ
Printed in Great Britain by J W Arrowsmith Ltd, Bristol

Contents

Preface

The study of extraterrestrial magnetic fields started in this century. By the end of the last century it was well established that the Earth had a magnetic field, and many of its properties had been investigated, including the fact that certain variations of the field were connected with the sunspot cycle. It had also been suggested, in the last century, on the basis of the shape of the corona during a total solar eclipse, that the Sun might also have a magnetic field. However, this suggestion had to await the observations of Hale and his co-workers before it was confirmed in 1908.

At first progress was slow. By 1947 the solar magnetic field had been studied in some detail and Babcock had also measured stellar magnetic fields. The presence of magnetic fields in interstellar space was first inferred from the polarisation of starlight, which was measured by Hall and Hiltner in 1949.

In the last 30 years progress has been much more rapid. This has largely been as a result of three separate but related factors. The first factor is the development of the new astronomies of radio, x-rays, infrared and ultraviolet. The second factor has been the use of space probes to investigate planetary and interplanetary environments. The third has been vast improvement in the design of instrumentation for optical telescopes. All these techniques have considerably increased our knowledge of magnetic fields in the universe. It is now clear that some planets, many stars and several galaxies have magnetic fields which play varying roles in the structure and evolution of these objects.

Cosmic magnetism is a subject of considerable breadth and its more technical aspects have been covered in many conference reports and a few specialised monographs. However, because of its mathematical nature, much of this material is not easily accessible to undergraduate students interested in astronomy and amateur astronomers. The subject can, however, be made accessible to a wider audience by extensive use of Faraday's very powerful and highly pictorial concept of 'lines of magnetic force' and their associated physical properties. It is this concept that is used throughout this book to explain the structure and behaviour of magnetic fields in extraterrestrial objects. It is my hope that this approach will make more people aware of the importance of magnetic fields in astronomy.

PAHS
Plymouth, February 1986

Chapter One

Introduction—Why Astronomy?

Why is astronomy important? Why should society fund fundamental research, and in particular why is it necessary to study cosmic magnetism? This chapter looks briefly at the changing relevance of astronomy to society and then examines the particular relevance of current astronomical research. Some of the most important contributions being made by astronomers to the modern world are looked at, and the place of magnetic fields in our current scheme of ideas is explained.

The Relevance of Astronomy

Astronomy arose independently in many parts of the world out of a practical human need for a calendar. Early communities of hunter-gatherers would have noticed, for example, that when certain patterns of stars were on the eastern horizon immediately after sunset, it was always at the time of year when certain berries were ripe for eating. They may also have noticed that some birds and animals would disappear from the landscape for a period, only to reappear with the changing seasons, when other patterns of stars were visible in the evening sky. In this way the sky would be quite useful as a calendar, and when eventually these early peoples settled down as farmers, the sky calendar helped them to decide when to sow seeds and when to prepare for the following seasons.

Ancient astronomers also used the Sun to tell time during the day and the stars to tell time at night. We know from the way they built their temples and pyramids that many early civilisations could find direction from the Sun and stars. The making of calendars, telling time and direction-finding were the most important tasks of the ancient astronomers.

Astrology provided the most important motive for studying the motions of the planets against the background stars. This remained so from the time of its origins in the ancient world right up to the period of the Copernican Revolution. Astrologers believed that human destiny was influenced by planetary motion, and that predicting the movements of the planets would enable them to predict future events on Earth.

The problems of finding latitude and longitude at sea provided the bridge between ancient and modern astronomy. Early requirements of astronomy did not require a very high level of accuracy. However, to find latitude and longitude with reasonable certainty required that the positions of the stars and the motion of the Moon be known with much greater accuracy than astronomers had previously been able to attain. This led to the establishment of the Paris and Greenwich Observatories in 1667 and 1675 respectively. The quantitative accuracy introduced into astronomy by these institutions laid the foundations of the modern quantitative approach to all astronomical measurements.

There are three main themes of astronomical research that are of direct relevance to modern society, as well as areas in which astronomy makes an indirect contribution by providing a service to other scientific subjects. First, astronomy is concerned with the structure, evolution and origin of the universe and its constituent parts. Second, it examines the forces, fields, radiation and particles that link Earth with the rest of the universe. Third, the universe is a gigantic, physical laboratory, where astronomical phenomena can be studied in ways that are simply not possible in the localised environment of Earth. In a general way astronomy makes contributions to mathematics, chemistry, biology and the geophysical sciences.

The relevance of astronomy has thus changed over the centuries. The main factors affecting this changing role have been the varying needs of society, technological progress and the general development of science.

The Structure and Evolution of the Universe

Thomas Kuhn in his book *The Copernican Revolution* says: 'Every civilization of which we have records has had an answer for the question "What is the structure of the universe?".' However, it is only in the Western civilisations which stem from Ancient Greece that astronomical observations have been used to arrive at an answer. In many cultures the answer was a concensus of social opinions based on religious beliefs, mythology or philosophical argument. In our civilisation we have delegated the task of finding an answer to this question to specialists in astronomical observations and theoretical physics. However, the world view that they produce influences many aspects of our lives from art and literature to philosophy and religion, so it is necessary that the most important results of astronomical research should be available to everyone.

In Western science we are not content merely to describe the size and structure of the universe: we want to understand the structure, evolution and origin of the cosmos in terms of the currently accepted laws of science. In order to answer these questions, we want to know how valid these laws are in space and time. We need to know about the origin of the chemical elements. We need to know about the evolution of stars and galaxies, because evolution in these objects can and does influence our observations on the large-scale structure of the universe. We also need to know how our observations are affected by our more local environment. We need to know about magnetic fields, because they can influence the structure and dynamics of celestial objects. Professor E N Parker in his book *Cosmical Magnetic Fields* says: 'The magnetic field exists in the universe as an "organism" feeding on the general energy flow from stars and galaxies. The presence of a weak field causes a small amount of energy to be diverted into generating more field, and that small diversion is responsible for the restless activity in the solar system, the galaxy and the universe.'

Earth and the Cosmic Environment

It is part of the role of the astronomer to assist in the task of understanding our terrestrial environment, and to determine the nature of the radiation, the forces and the fields whereby the extraterrestrial universe can influence Earth. We know that Earth is bathed in a wide range of electromagnetic radiation. The motion of the Earth around the Sun and the tides of the oceans are controlled by the forces of gravitation. We are continually being bombarded by energetic particles from the rest of the universe. Magnetic fields are yet another link between Earth and the rest of space.

Astronomy also makes a less direct contribution to the geophysical sciences. Since we cannot alter to any significant extent the basic properties of Earth, it is useful to have alternative environments in which to test the consequences of certain geophysical theories. The geology of the planets is teaching us more about the geology of Earth. The meteorology of planetary atmospheres is shedding some light on our understanding of our own atmosphere. The study of planetary magnetic fields and magnetospheres is helping us to understand the geomagnetic field and our own magnetosphere.

There is growing evidence for links between life and the geomagnetic field, and some biologists believe that the effects of Earth's field on life might well be more important than has been realised up to now. These biologists believe it is important for us to understand the behaviour of this field and to understand solar–terrestrial relationships.

Celestial Physical Laboratories

'The Case for Astronomy' was argued by the Nobel prize-winning physicist S Chandrasekhar in a lecture which he gave to the American Philosophical Society several years ago. In this lecture he quoted four specific examples where astrophysical conditions provided the first tests for physical theories.

The first example concerns the structure of white dwarf stars and the special physical laws applicable to them. As will be seen in a later chapter, these stars have a density of matter very much higher than thought possible under the normal laws of physics, as they were appreciated at the time of their discovery. However, physicists were just beginning to understand that, when dealing with subatomic phenomena, a new set of rules applied—the rules of quantum mechanics. When dealing with large numbers of subatomic particles such as electrons these rules took a particular form known as quantum statistical mechanics. When applied to the structure of white dwarf stars quantum statistical mechanics was able to give an explanation for why these stars did not collapse under their own gravitational forces at such high densities. This was the first application of this particular branch of physics. However, this theory was later applied to the understanding of the structure of metals and related problems in solid state physics.

The second example is related to the dark appearance of the Sun's visible disc—often seen in photographs of the Sun—near the circumference. This phenomenon is known as limb darkening. Physicists were able to explain this in terms of the presence of negative hydrogen ions (i.e. hydrogen atoms with an extra electron) in the photosphere of the Sun. This was before this ion had been observed under laboratory conditions.

Understanding the energy generation in stars provides a third example. The source of energy for the Sun and stars had long been a problem for physicists and astronomers. At one stage it was believed that the stars were fuelled by their gravitational collapse from an enormous cloud of gas. As the particles of this gas fell in towards the centre of the cloud they would have gained energy, and were thus moving very fast when they eventually formed the gaseous spheres that we call stars. This line of argument, however, led to an age for the Sun which was less than that of the Earth as estimated by geologists. The difficulty was overcome when it was realised that nuclear reactions were taking place in the interiors of stars. This was the first time that the understanding of nuclear reactions had been applied to any physical situation.

Synchrotron radiation is the type of radiation emitted by large atom smashers as high energy electrons spiral around the field lines of these machines. However, the theory of synchrotron radiation was worked out, as an academic exercise, long before atom smashers were invented. It received its first application to the astronomical problem of explaining the source of energy of the Crab nebula. This was the fourth example quoted by Chandrasekhar.

There are other examples of great physical theories being developed in astronomical contexts. Newton's work on mechanics and gravitation was inspired by his attempts to understand planetary motion and to reconcile the laws of physics with a moving Earth. The accepted cosmology, before the Copernican Revolution, was that based on the work of the Greek philosopher, Aristotle. Aristotelian cosmology was not only a scheme for how the celestial bodies were arranged in space, but also included doctrine on the laws of physics. This cosmology taught that two different sets of laws were in operation: one set was applicable to the region below the Moon, which included the Earth; the other set of laws applied to the region above the Moon, the celestial region. Copernicus' suggested rearrangement of Aristotle's scheme for the celestial bodies, with the Sun rather than the Earth at the centre, was not consistent with Aristotelian physics. Although Copernicus, as well as some of his contemporaries and successors, tried to reconcile the astronomy with the physics, it was left to Newton to build on the work of his predecessors and produce a grand synthesis. Newton's laws of motion and his law of gravitation were universal, so not only did they explain celestial motions, they also applied to earthly mechanics. Celestial situations thus provided the stimulus as well as the severest testing ground for Newtonian mechanics. This was because the near vacuum and frictionless environment of the planets could not, at that stage in the history of physics, be matched in the laboratory. The celestial 'laboratory' provided the opportunity to concentrate on the essentials of dynamics.

The speed of light is, of course, an important physical constant. It plays a very important and fundamental role in Maxwell's theory of electricity, magnetism and light, and in Einstein's work on relativity.

Yet the fact that light did have a finite and measurable speed was not realised until Roemer made this discovery in connection with his research on the moons of Jupiter. He realised that his predicted times for the eclipses of Jupiter's moons were more and more in error as the distance between Jupiter and the Earth increased. He later attributed this to the fact that the light, which carried the information on eclipses, had a greater distance to travel the further Jupiter moved from Earth, and therefore must move at a finite speed. Closely related to this phenomenon was Bradley's discovery of the aberration of starlight. A star is never really in the place we seem to see it, for several reasons, one being due to aberration. Because light has a finite speed, and because the Earth is moving around the Sun, the telescope with which we are observing a star must be tilted with respect to the actual direction of the star. Many years later George Biddell Airy, seventh Astronomer Royal at Greenwich, investigated the effect of a water filled telescope on aberration. The negative result of this experiment—i.e. that aberration was not affected by the presence of the water, was later explained in a convincing way by Einstein's special theory of relativity.

Much of our knowledge of the external universe is based on the spectroscopic analysis of celestial objects. Although spectroscopy has been just as important for chemists and physicists, astronomy has played a vital role at important stages in the development of this branch of science. In his inaugural lecture as Jacksonian Professor of Natural Philosophy at Cambridge, Professor A H Cook drew attention to the contribution made by astronomy to several spectroscopic problems. His first example was the discovery of helium. Sir Norman Lockyer discovered this element in the atmosphere of the Sun by his careful studies of the solar spectrum some years before helium was detected in the laboratory. Some important techniques and data, very necessary for the spectroscopic identification of chemical elements, were worked out in astronomical contexts before they were used by chemists. Professor Cook's third example was the Balmer series of spectral lines of hydrogen. Although this series had previously been theoretically predicted, it was first verified in the spectra of stars. The nature of the so-called 'forbidden lines' in the spectra of planetary nebulae and the solar corona provided his last example. These lines were called forbidden because, according to the rules of the old quantum theory, the atomic transitions (i.e. the rearrangements which take place in the atom) which would give rise to the lines, were forbidden. According to the new quantum mechanics these lines are not really forbidden: it is just that they arise from atomic states that have long lifetimes—the transitions involving these states thus have low probabilities, and therefore are difficult or impossible to detect under laboratory conditions.

From these examples Professor Cook concluded, '. . . astronomical observations have been crucial in the development of atomic spectroscopy, and I believe that the reason is that astronomical sources afforded conditions which were not at that time accessible in laboratory sources'.

The particular problem of forbidden lines had another important effect on the history of physics. A physicist called Nicholson published a series of papers arguing that the most primitive forms of matter existed in objects such as gaseous nebulae and the solar corona. He believed the atoms of special elements in these objects consisted simply of a ring of a few electrons surrounding a positive but very small nucleus. He also applied the ideas of early quantum theory to these hypothetical atoms and was able to account for some of the spectral lines in the solar corona and in nebulae. This was before Rutherford conceived his own ideas on atomic structure. Bohr's work on the subject built on Rutherford's ideas, but his first papers on the subject acknowledge, correct and extend the pioneering work of Nicholson.

Plasma physics is concerned with the study of the interaction between extremely hot gases (i.e. gases in which most of the atoms are ionised) and magnetic fields. It is a subject of intense interest to many terrestrial physicists because the harnessing of solar-type nuclear reactions to our energy needs vitally depends on understanding plasmas and their containment by magnetic fields. It is, moreover, a subject to which astronomy has made many contributions—the universe provides a series of plasma physics laboratories of varying dimensions. In 1970 the Nobel Prize for Physics went to Hannes Alfven 'for fundamental work and discoveries in magnetohydrodynamics and their fruitful application in different parts of plasma physics'. Much of his work was inspired and motivated by astrophysical situations. In 1942 he discovered a special type of plasma (or hydromagnetic) wave, now called Alfven waves, in the ionised gases of the Sun while he was researching into sunspots. In astrophysical situations magnetic fields and plasma exhibit behaviour that cannot always be reproduced in laboratory situations, but which can increase our understanding of plasma physics.

Astronomy and Mathematics

Astronomy has also stimulated research and discovery in those areas of mathematics which are often used by applied mathematicians and physicists. The great French mathematician Fourier once said, 'The profound study of nature is the most fecund source of mathematical discovery.' In the ancient world many problems in geometry, plane and spherical trigonometry first arose in astronomy. Newton's invention and development of calculus was motivated largely by his work on planetary motion and the problem of calculating the gravitational field of a spherical body. A few important mathematical functions and techniques were invented by French and German mathematicians working on the motions of planets and the gravitational fields of non-spherical bodies. A special method of approximation was invented by a Swiss mathematician in his efforts to solve the problem of the Moon's motion.

Most of the mathematical problems posed by astrophysical plasmas moving in magnetic fields are so difficult that they can only be solved by making a number of special approximations. The full treatment of most of these problems lies outside the range of our mathematical techniques at the moment, but some of the problems could well stimulate further research into certain areas of mathematics.

Summary

It is important to study cosmic magnetic fields if we are to understand the structure, evolution and origin of our universe; to understand the links between our terrestrial and extraterrestrial environments; and to learn more about the interaction between plasmas and magnetic fields. This last point has been put very well by Professor Parker: 'Over astronomical dimensions the magnetic field takes on qualitative characteristics that are unknown in the terrestrial laboratory. The cosmos becomes the laboratory, then, in which to discover and understand the magnetic field and to apprehend its consequences.'

Chapter Two

The Forces of Nature

Poets, artists and scientists have commented on, investigated and tried to capture the variety and complexity of the physical universe from the very small scale to the incomprehensibly large. So perhaps it is surprising that much of this complex diversity can successfully be understood in terms of four basic physical forces. This chapter outlines briefly the nature of the four forces, the roles that three of these forces play in astronomy, and, in more detail, the properties of the electromagnetic force.

The Forces of Physics

The four fundamental forces of physics are: the strong nuclear force; the weak nuclear force; gravitation; and the electromagnetic force. The nuclear forces are very short in range and only act over distances comparable in size with the nucleus of an atom. Gravitation is the weakest of all the forces, but it acts between all bodies that have mass, and it is long range. The electromagnetic force acts at a variety of levels from the very small scale to the extremely large scale, but it is most important to physicists, chemists and engineers—i.e. it plays an essential part in those aspects of the physical universe that are encountered in our local environment and everyday life.

Nuclear Forces and Astronomy

The nucleus of an atom contains protons and neutrons. Since the protons carry positive charge, and like charges repel each other, there must be an extremely strong force binding the nucleus together which can overcome this repulsion. This is the strong nuclear force. Physicists now believe that neutrons and protons are made up of yet smaller particles called quarks, and that the strong nuclear force is a remnant of a much more powerful force, called the colour force, acting on quarks inside the protons and neutrons.

The weak nuclear force operates when a neutron decays into a proton, emitting an electron and a neutrino. It thus plays a part in the radioactive decay of many nuclei. The most important role for nuclear forces in astronomy is in providing the source of energy in the deep interiors of stars. Here the temperatures are so high that the thermal energies of nuclei are high enough to overcome the electrostatic repulsion that exists between them. This means that nuclear reactions can take place in which the fusion of two nuclei can occur with the consequent release of enormous amounts of energy. This process is not only responsible for fuelling the stars, it is also partly responsible for the manufacture of heavier chemical elements from lighter ones. Nuclear reactions, leading to the synthesis of elements, also take place in the region of red giant stars, in novae and supernovae. Because nuclear forces are short range they cannot directly influence the area outside the reacting core of a star, but they can do so indirectly by means of particles (such as neutrinos) and electromagnetic radiation, which are produced as a result of nuclear reactions.

Gravitation and Astronomy

Gravitation is an interaction which occurs between every pair of particles in the universe. According to Newton's law of gravity the force between any two particles is directly proportional to the product of their masses, and inversely proportional to the square of the distance between them. The one exception to this rule is the neutrino, a subatomic particle associated with some types of radioactive decay. Even the photon—the particle associated with electromagnetic radiation—is subject to the gravitational interaction. Newton's law implies that the force of attraction between two bodies gets weaker as the distance between them is increased. The weakening of the force by distance means that in most cases the gravitational influences of very distant objects can be ignored, even if they are massive. So, when we stand on Earth, its own gravitational field is by far the most dominant one we experience. However, in discussing the tides of the ocean we must consider the gravitational pull of the Moon and the Sun. Because the Moon is nearer to the Earth than the Sun is, its influence on the tides is greater than that of the Sun, although the Sun is of course far more massive than the Moon.

The shapes of planets are mostly determined by the force of gravity, modified slightly by rotation. The maximum height of a mountain or the depth of a valley is determined by the strength of the materials of the Earth's crust and the surface gravity of our planet, and the same is true for the other terrestrial planets. The atmospheres of those planets which have atmospheres are 'contained' by the gravitational fields of the planets. On a larger scale the motions of the planets around the Sun and the orbits of satellites are completely controlled by gravitational forces.

The equilibrium structures of our Sun and other stable stars are the result of a delicate balance between pressure forces and the force of gravitation at each point within the stars. Galaxies are large collections of stars bound together by gravitation, and clusters of galaxies are also held together by their mutual gravitational attraction. The dynamics of the universe as a whole is a consequence of gravitation, although on this extremely large scale Einstein's general theory of relativity becomes more appropriate than Newton's law.

Since gravity plays so large a part in the structure and dynamics of the universe, what is the role of magnetic fields? Let us look first at the nature of electromagnetic forces.

Magnets and Magnetic Fields

Most of us are familiar with the basic properties of magnets, either from recollections of school science lessons or through the use of magnets at home or at work. Probably the best known property of a magnet is the ability to attract certain materials without any physical contact, and this attractive influence can penetrate certain other materials and substances. For instance, magnetic earrings are kept in place by a small magnet placed behind the ear, so the attractive force of the magnet is capable of penetrating the live tissue of the earlobe, without of course harming it in any way. This region of influence surrounding the magnet is called the magnetic field.

A good way to show the existence and extent of this field is to sprinkle iron filings on a sheet of cardboard placed on top of a bar magnet. When the card is tapped lightly the iron filings will be arranged by the magnetic field into a clear pattern. This shows that the magnetic field acts along lines, and the iron filings will align themselves along these 'lines of force'. The lines seem to radiate from the ends of the bar magnet and join together half way along the magnet. If a small compass is brought close to one end of the magnet, the compass needle will point directly away from it, whereas if brought close to the other end the needle will point directly towards the magnet (see figure 2.1). There appear to be two points near the ends of the magnet from which the lines radiate: these points are called the poles of the magnet. The poles can be found by placing a small compass at several points around the end of the magnet, marking the directions in which the needle points at each position, and drawing straight lines through these directions. The point of intersection of the straight lines defines the pole at each end of the magnet (figure 2.2).

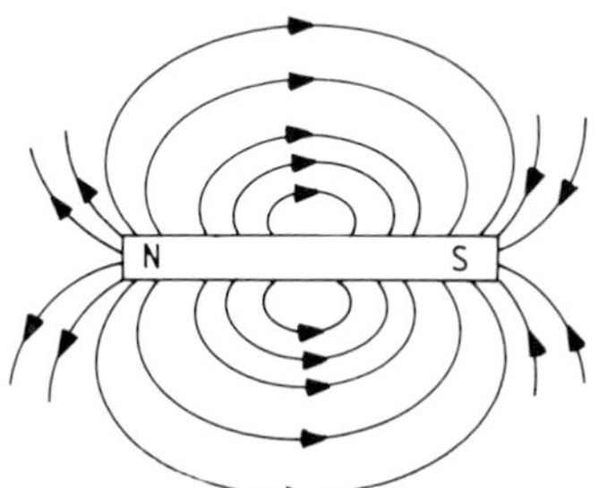

Figure 2.1 Magnetic lines of force around a bar magnet.

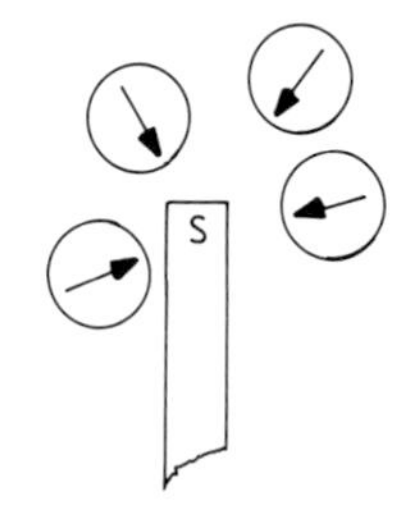

Figure 2.2 Small search compasses pointing towards the pole of a bar magnet.

The magnetic lines of force are intangible lines along which a compass needle will point or along which iron filings will align themselves if free to move under the influence of the magnet. Although these lines should not be thought of as real physical entities, they are an extremely useful illustrative concept. Michael Faraday, the great English scientist who lived from 1791 to 1867, was able to explain a great deal by giving the 'lines of force' real physical properties. He believed that the lines of force were something like elastic bands in that they tended to contract along their lengths. However, if there were a bundle of such lines all pointing in the same direction, then the lines would tend to repel each other at right angles to the direction in which they were pointing. In other words there is a tension along their lengths and a pressure at right angles to their lengths. The properties of magnets can be explained in terms of the interactions between these lines of force.

Suppose two bar magnets are marked with red paint at the ends to which a small compass needle points directly. If the two red-painted ends of the magnets are brought up to each other they repel each other. However, if the painted end of one magnet is brought up to the unpainted end of the other magnet, they attract each other and join together. This shows that unlike poles attract each other and like poles repel each other. It is also possible to investigate the magnetic field surrounding these magnets when they are placed in different positions with respect to each other (figure 2.3). The magnets are stuck to a table top with sticky tape, and a sheet of card placed over them. Iron filings are sprinkled onto the card which is then gently tapped. The configurations taken up by the iron filings will be as indicated in figure 2.3. The directions in which the lines of force are pointing can be found using a small compass. In figures 2.3(*a*) and (*b*) the lines of force between the magnets are pointing in the same direction, so the magnetic pressure between the lines will be transferred to the magnets to which they are attached: hence the magnets will repel each other. In cases (*c*) and (*d*) the lines of force will tend to contract along their lengths, so the magnets will tend to attract each other. The explanation of forces between magnets in terms of lines of force surrounding them, illustrates how powerful the concept of these lines is when discussing magnetism.

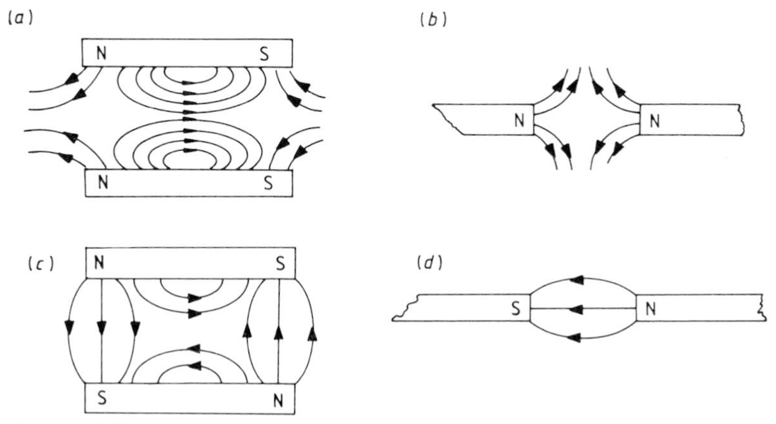

Figure 2.3 (*a*) Lines of force around two bar magnets with like poles opposite each other. (*b*) Lines of force around the facing like poles of two bar magnets. (*c*) Lines of force around two bar magnets with unlike poles opposite each other. (*d*) Lines of force around the unlike poles of two bar magnets.

If a bar magnet were to be broken into very small pieces, and each piece tested with a compass, it would be found that each piece behaved as if it were a complete magnet in itself. It is now known that a magnet consists of a large number of very tiny regions, called domains, each of which is like a little magnet. In an unmagnetised piece of iron these domains are randomly orientated, but in a magnetised piece of iron the domains are aligned so that they all point in the same direction (see figure 2.4). This alignment can be achieved by stroking an unmagnetised piece of iron with a magnet (figure 2.4(*b*)). If a magnet is heated above a certain temperature (known as the Curie point) it will cease to exhibit magnetic properties, and can only be remagnetised once it has cooled below the Curie point. This can be explained in terms of the domain theory of magnetism. As the magnet is heated up the domains tend to move about more violently, and the violence of the movements will increase with temperature. Eventually a temperature will be reached where the domains move about so violently that they no longer sustain their original alignment, and the magnetism of the object is lost. If this unmagnetised iron is then allowed to cool in a magnetic field, for example the Earth's field, then at least some of the domains will become aligned in the field and the iron will become slightly magnetised in the direction of the external field. The number of domains that will become aligned in the external field will depend on the strength of that field, hence the strength of the resulting magnet will be a measure of the strength of the field in which it was magnetised. This fact has been very useful in investigating the history of the Earth's magnetic field, and the application of this technique to the history of the Earth's field is discussed in more detail in the next chapter.

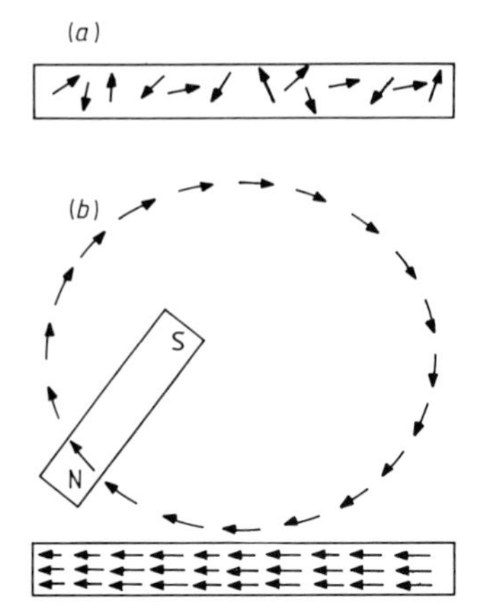

Figure 2.4 (*a*) Unaligned domains in a soft iron bar. (*b*) Aligning the domains by stroking with a bar magnet.

Electricity and Magnetism

Magnetic lines of force or magnetic fields can be generated by the flow of electric current. If an electric current is passed through a coil of wire this will give rise to a magnetic field surrounding the coil. If this field is investigated using a small compass it will be noticed that it appears very similar to the magnetic field of a bar magnet (see figure 2.5). The lines of force form complete loops which thread their way through the interior of the coil. According to the modern theory of magnetic materials it is now believed that the field lines of bar magnets also thread their way through the interiors of these magnets and that the field itself is generated by electric currents within the atoms of the magnets.

A straight wire carrying an electric current produces a magnetic field in which the field lines take the form of concentric circles centred on the wire (see figure 2.6). Two parallel wires each carrying a current in the same direction will be attracted to each other. This can also be explained in terms of the lines of force of their combined fields. The circular lines of each individual field will cancel in the region between the wires because they are in opposite directions. There will then be a reconnection of the lines forming the rest of the circles and the lines in the resulting loop will tend to contract along their lengths, giving rise to an attraction between the wires (figure 2.7). The process of reconnection will be seen to be important when discussing the behaviour of field lines on the Sun.

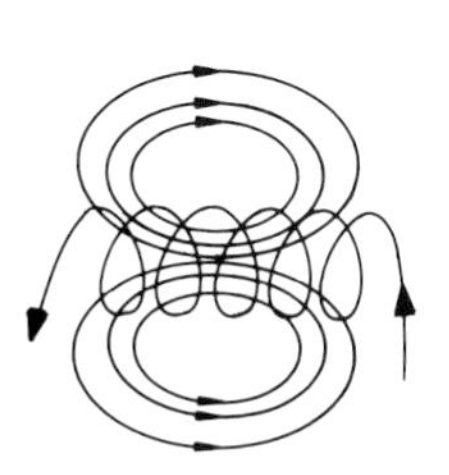

Figure 2.5 Lines of force around a coil of wire.

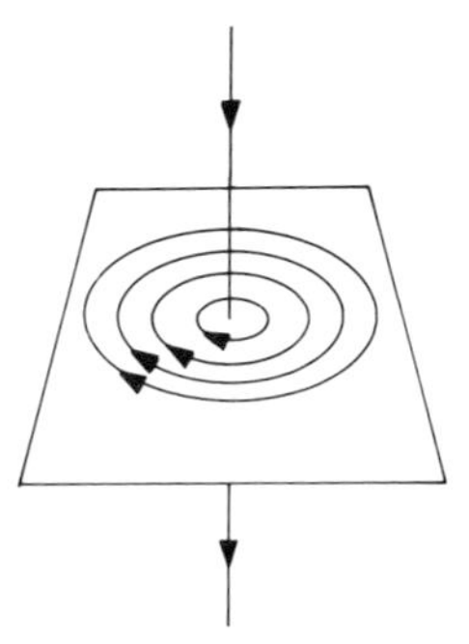

Figure 2.6 Magnetic field around a straight conductor.

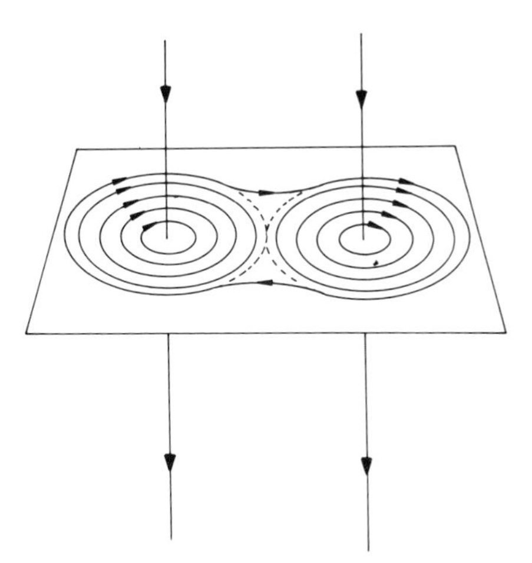

Figure 2.7 Reconnection of lines of force around two straight conductors.

Faraday also made the discovery that moving magnetic fields can produce electric currents. If a coil of wire is connected to a device for measuring electric current, called an ammeter, and a magnet is moved in and out of the coil, the ammeter registers a fluctuating current (see figure 2.8). In this experiment it does not really matter if the magnet is moved or the coil is moved—in either case an electric current is generated. This principle is the basis of the electric generator.

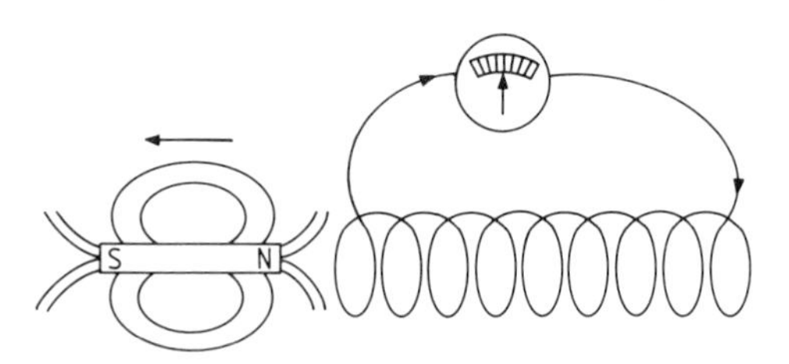

Figure 2.8 Generating electric current by moving a magnet with respect to a coil of wire.

The last three demonstrations have shown how electricity and magnetic fields are intimately linked. Maxwell was able to give a brilliant expression of this in his electromagnetic theory, and Einstein was able to explain this in terms of the special theory of relativity.

A charged particle moving in a magnetic field experiences a force which is at right angles to the direction of its motion, and at right angles to the magnetic field (figure 2.9). If the field is extensive enough and uniform in strength, and the particle's velocity is at right angles to the field lines, then its path will be a circle about the lines. If the particle's velocity makes an angle with the field lines other than a right angle,

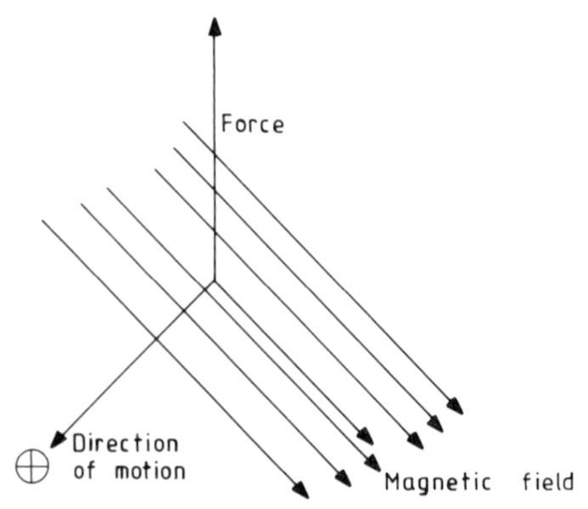

Figure 2.9 The force acting on a charged particle moving in a magnetic field.

then its path will be a helix about the lines (figure 2.10). The principle of an electron being deflected by a magnetic field is used extensively in much modern equipment, for example in the scanning mechanism of a television set, or in an electron microscope.

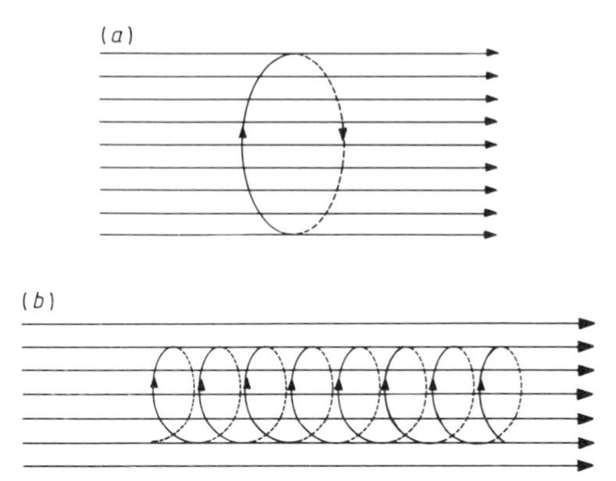

Figure 2.10 (*a*) Orbit of particle with motion at right angles to lines of force. (*b*) Orbit of particle with direction of motion making an angle of less than 90° with field lines.

A wire carrying an electric current consists of charges in motion within the wire, so these charges will experience a mechanical force if the wire is placed in a magnetic field. In particular a small coil placed in a magnetic field has a field of its own, which is similar to that of a bar magnet, when a current is passed through it, so it will tend to orientate itself with the plane of the coil at right angles to the external field. This principle is used in electrical measuring equipment and forms the basis of all electric motors.

Electromagnetic Radiation

So far we have looked at the close relationship between electricity and magnetism. Now let us look at how these phenomena are linked with light and radio waves. Faraday's experiments had shown that such links existed, but it needed the mathematical genius of the physicist James Clerk Maxwell to combine the results of Faraday's experiments and his concept of lines of force to produce a theory which explained light, radio waves, x-rays, infrared and ultraviolet radiation, in terms of electrical and magnetic fluctuations. Radio waves can be generated by passing a rapidly fluctuating electric current through a straight conductor, and this is basically how radio and television transmitting aerials work. The wave coming from such a conductor would have an electrically fluctuating part parallel to the conductor and a magnetically varying part at right angles to the conductor (figure 2.11). In most circumstances it is the electrical part of the wave that is detected, so the wave is normally described as being polarised parallel to the conductor. The wave transmitted by such an aerial can be received by an aerial consisting of a straight conductor parallel to the length of the transmitting aerial. Some radiation will also be received if the two aerials make an angle of less than 90° with respect to each other, but hardly any radiation will be received if the angle between the aerials is 90°.

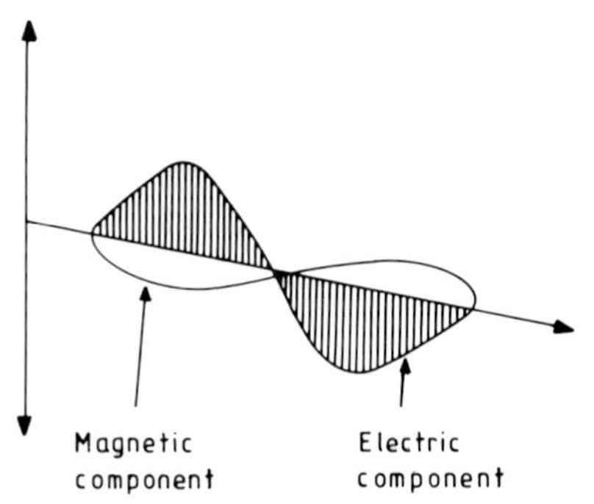

Figure 2.11 The electric and magnetic components of a plane-polarised electromagnetic wave.

Light waves cannot be generated in this way, but electrons within the atom can generate a variety of different wavelengths from x-rays to infrared. In explaining how this is achieved it is helpful to start with a very simple picture of the hydrogen atom devised by Niels Bohr. Just as the force of gravity of the Sun on a planet is inversely proportional to the square of the distance between them, so the force—due to electrostatic attraction—between a proton and an electron obeys the same law. In the Bohr model the electrons are orbiting the central nucleus, in much the same way as planets orbit the Sun, but here the force is provided by the electrostatic field rather than by gravity. Bohr's model of the hydrogen atom consists of one electron moving in a circular orbit about the positively charged nucleus, which in this case is a single proton. According to the laws of classical physics the electron would radiate electromagnetic waves as it orbits, lose energy in the process, and thus spiral into the nucleus. To avoid this, Bohr applied the ideas of quantum theory to the orbiting electron, and showed that the electron could only move in orbits which were at certain 'allowed' distances from the nucleus (figure 2.12). These allowed orbits are more easily explained in terms of the full quantum theory, and an explanation involving this branch of modern physics is beyond the scope of this book. However, it is possible to give a simplified explanation in terms of the ideas developed by Louis de Broglie. He assumed that associated with any moving particle there was a wavelength, called the de Broglie wavelength. This depends on the speed and mass of the particle. The allowed Bohr orbits then correspond to those in which the speed of the electron is such that the associated de Broglie wavelengths just fit a whole number of times into the circumference of the orbit (see figure 2.13).

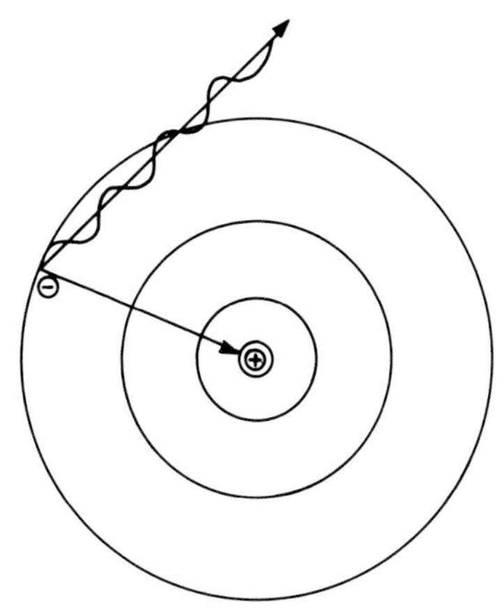

Figure 2.12 A diagrammatic view of Bohr's model of the hydrogen atom.

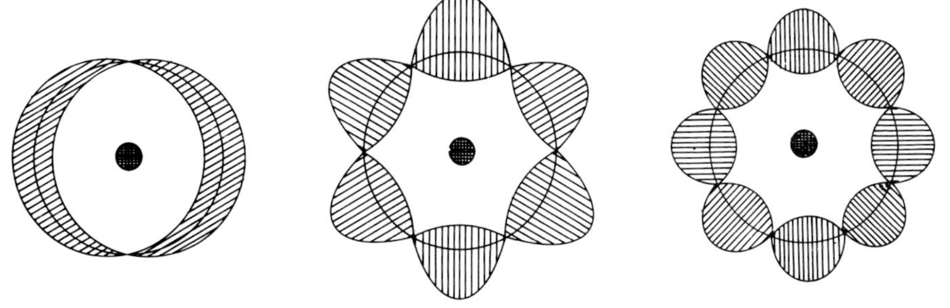

Figure 2.13 De Broglie waves and the allowed states of the Bohr atom.

When the atom absorbs light (or any other electromagnetic waves) the electron is shifted from an orbit close to the nucleus to one further away. Later the electron will 'fall' from its 'excited' state, and when this happens the atom will broadcast light at a very precise wavelength. In other words it will emit a very narrow band of colour, visible with a spectroscope, rather than the entire rainbow.

A move from each orbit far from the nucleus to one closer to the nucleus will give rise to a different wavelength, so the atom will emit a well defined series of wavelengths that is characteristic of that particular type of atom. The heavier chemical elements have more complex atoms. The nuclei of these atoms consist of protons and neutrons, and a number of electrons (equal to the number of protons in the nucleus) orbiting the nucleus. Each type of atom will emit its own set pattern of definite wavelengths. It is this distinctive set of wavelengths transmitted by the atoms of different elements that enables the chemical elements to be identified by the radiation emitted. The branch of physics dealing with this aspect of radiation is called spectroscopy, and it enables the chemical composition of stellar atmospheres to be deduced from an analysis of their spectra. The spectrum of a star consists of the continuous spectrum crossed by a number of dark absorption lines. The continuous spectrum arises from the photosphere of the star, whereas the dark lines are due to the absorption of light by the atoms of various gases in the atmosphere of the star. These gases can be identified by comparing the positions of the dark lines with bright emission lines generated by known elements in the laboratory (figure 2.14). If the star is moving away from, or towards, Earth it will be necessary to 'move' the spectrum either towards the red end or towards the blue end, of the spectrum. This is necessary because of a physical principle known as the Doppler effect.

The Doppler effect was discovered in 1842 by the Austrian physicist, Christian Doppler, in connection with his work on sound waves. One of the best known examples is the sound of a whistle on a fast moving train. If you are standing still on a platform as a train goes by, you first hear

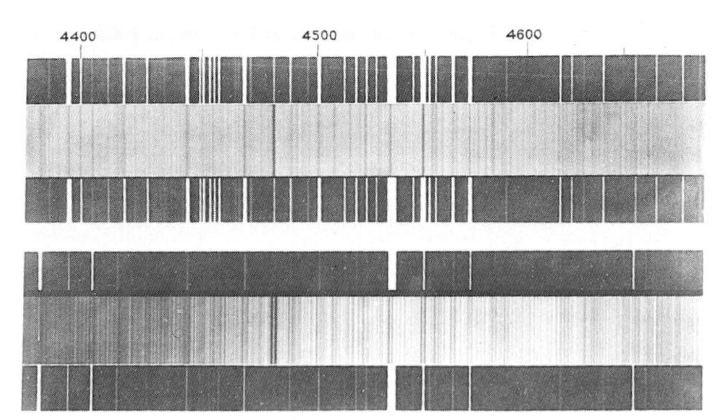

Figure 2.14 Spectra of a binary system. Above and below each of the spectra are comparison lines. In the lower spectrum the lines are split by the Doppler effect since one component is travelling towards and the other away from the observer. (Courtesy of the Royal Astronomical Society.)

the train whistle as a high-pitched sound as the train approaches. At the moment the whistle passes you, the sound suddenly changes and you hear a lower-pitched sound as the train rushes on its way. This is because sound consists of waves, which can be pictured as a series of peaks and troughs. The pitch of a given sound depends on how many peaks and troughs hit the ear in a given time. When the source of the sound waves is travelling towards us the number of peaks striking the ear in a given time is increased, and when the source is travelling away from us the number of peaks reaching the ear in the given amount of time is decreased. The train whistle travelling fast towards your ear as you stand on the platform sends a high number of peaks towards you, and then as it passes the number of peaks is suddenly reduced—this is what causes the change in pitch (figure 2.15).

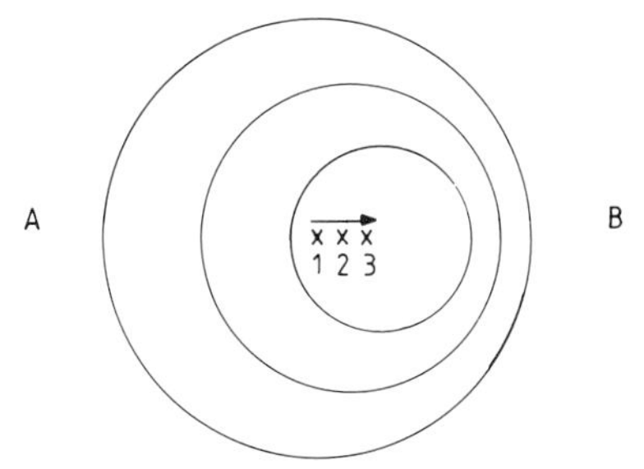

Figure 2.15 Diagram illustrating the Doppler effect. Numbers 1 to 3 are successive positions of a source of waves moving to the right.

Light waves can also be thought of as a series of peaks and troughs, and they can cause analogous effects. If a star is travelling towards us the dark absorption lines in its spectrum will be shifted towards the blue end of the spectrum, and if the star is moving away the lines will be shifted towards the red end of the spectrum. The amount of the shift tells us the speed of approach or recession of a star.

The speed of all the different types of electromagnetic radiation is the same in a vacuum, but in media, such as glass, air or water, the speed of the various types of radiation will depend on their wavelengths. A medium which affects the speed of electromagnetic radiation in this way is called a dispersive medium and the phenomenon is known as dispersion. In a later chapter we will see how dispersion can be used to find the distances to pulsars.

Normally the radiation emitted by atoms is not polarised except when the atoms are in a magnetic field. However, the light from an unpolarised source can be polarised by passing it through a polarising filter, for example one of the lenses of a pair of Polaroid sunglasses. It is possible to arrange two polarising filters in such a way that no light emerges through them. This happens when the plane of polarisation of one filter is at right angles to the plane of polarisation of the other filter.

Measuring Celestial Magnetic Fields

A physical principle known as the Zeeman effect provides astronomers with an important tool for investigating extraterrestrial magnetic fields. In the simplest case, if the atoms of an element are placed in a magnetic field each of the original spectral lines will be split into two components when observed along the field lines, and into three components when viewed at right angles to the magnetic lines of force. The two components seen along the field lines will be right- and left-hand circularly polarised. An ordinary wave can be visualised by a rope just moving up or down. Right- and left-hand circular polarisation can be visualised by a rope given either a clockwise or an anti-clockwise spiralling motion (see figure 2.16 (*b*), (*c*) and (*d*)). The three components seen at right angles to the field lines will be linearly polarised (see figure 2.16(*a*)). The wavelength differences between the components can be used to measure the strength of the magnetic field.

The synchrotron effect is another important principle used for investigating magnetic fields in astronomy. This effect is the result of electromagnetic radiation generated by electrons, travelling at speeds close to the speed of light, spiralling around magnetic lines of force. This effect can readily be understood in qualitative terms. According to Maxwell's electromagnetic theory electric lines of force can only start and end in charged particles—they cannot originate from points in empty space. If a charged particle is at rest, an observer some distance away from the particle will see the lines of force radiating out from the point where the particle is situated. If the particle moves from point A to point B (see figure 2.17) in a given time t, then the 'message' that the particle has moved can only reach the observer with the speed of light c. So if the observer is very far from the particle he will still think that the particle is at A, whereas an observer whose distance from the particle is less than ct will 'see' it at the new position B. On a sphere of radius ct the lines of force seen by the two observers must join up, so there will be 'kinks' in the lines on this sphere. These 'kinks' are the electromagnetic waves radiating out from the particle that has moved from A to B. Any changes in the motion of the particle, whether changes in speed or direction, can only be propagated to an observer at a speed equal to the speed of light. This means that any particle undergoing acceleration, i.e. changes in speed or direction, will have 'kinks' in its field lines which

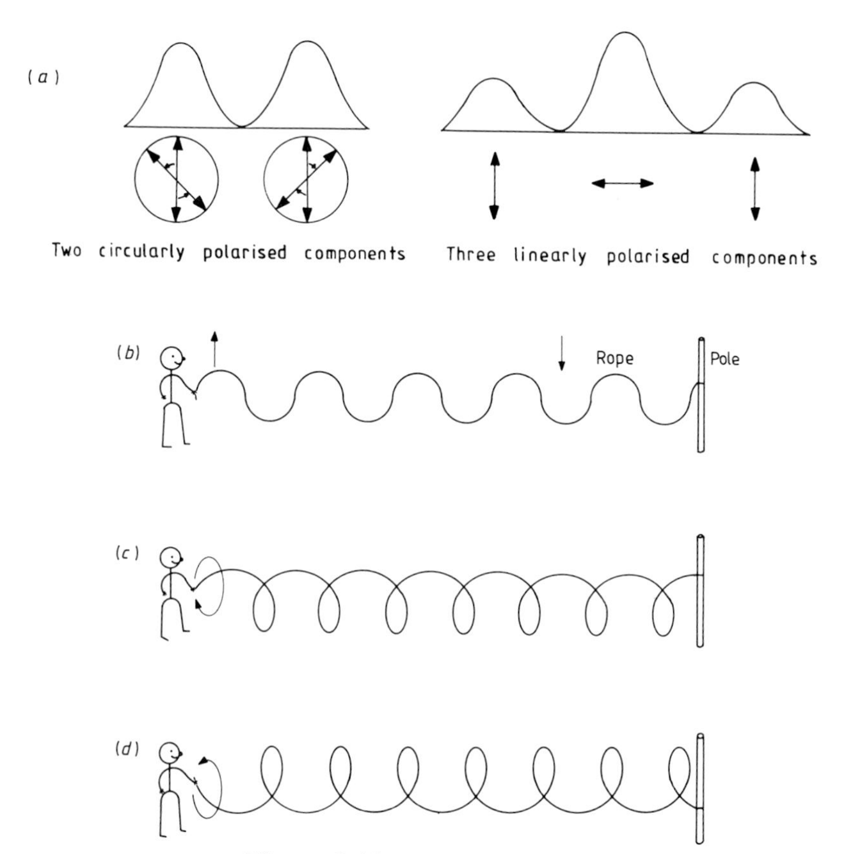

Figure 2.16 The Zeeman effect.

means that it will radiate electromagnetic waves. If the speed of the particle is small compared with the speed of light the intensity of the radiation varies with direction as compared with the direction in which the particle is moving (see figure 2.18). If, however, the speed of the particle is very close to the speed of light then most of the radiation will be concentrated in the region immediately in front of the direction in which the particle is moving, rather like the headlight of a motorcycle. The radiation is also polarised at right angles to the field lines if the particle in question is spiralling around magnetic lines of force. The intensity of radiation from a single particle is related to the strength of the magnetic field and to the energy of the particle, so measurements of the synchrotron effect can be used to investigate the structure and strength of magnetic fields.

The Faraday effect has also proved useful in studying magnetic fields in astronomy. According to this principle, when a plane polarised beam of light passes through a region containing electrons and a magnetic field, the plane of polarisation will be rotated about an axis which is the actual direction of the beam. The angle through which it is rotated depends on the distance over which the waves interact with the charged particles of the magnetic field, the angle between the line of sight and the magnetic field, the number of electrons per unit volume of space, and the strength of the magnetic field (see figure 2.19). When looking at right angles to the magnetic field there is no Faraday rotation to consider, but when looking along the field lines the Faraday rotation is a maximum. The strength of the synchrotron radiation is a maximum when observations are being made at right angles to the magnetic lines of force, and these observations are unaffected by Faraday rotation. The Faraday effect can be used to give information on the strength and direction of the field.

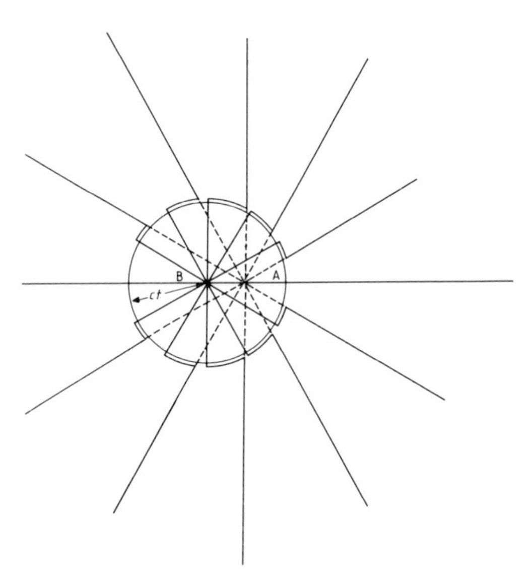

Figure 2.17 The lines of electric force radiating out from a moving point charge.

The magnetic fields of the Earth and planets can be measured using magnetometers on board satellites and space probes. There are two basic types of magnetometer that have been used for this purpose. The first type is called a flux gate magnetometer. In its basic form it consists of a core of magnetic material around which two conducting coils are wound, and in principle it is similar to an ordinary transformer. The coils are insulated from each other and from the core. A varying electric current of known frequency is passed through one coil and the current induced in the other coil is analysed by special frequency detectors. In

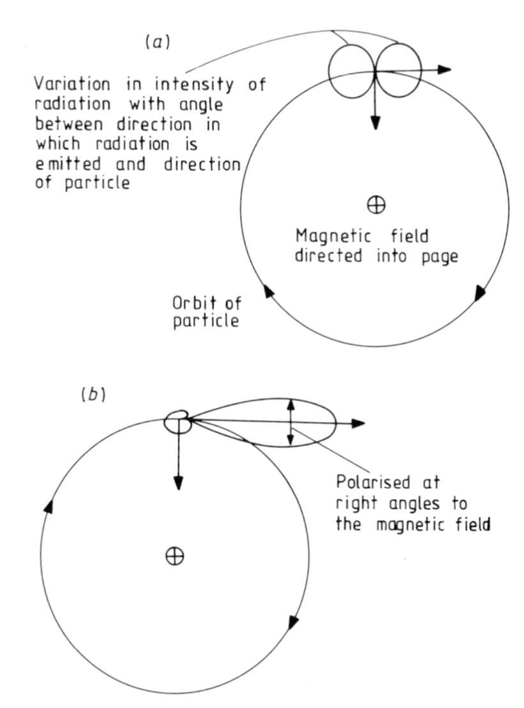

Figure 2.18 (*a*) Radiation from a particle moving in a magnetic field for the case when the speed of the particle is small compared with the speed of light. (*b*) Radiation from a particle moving in a magnetic field for the case when the speed of the particle is approaching the speed of light.

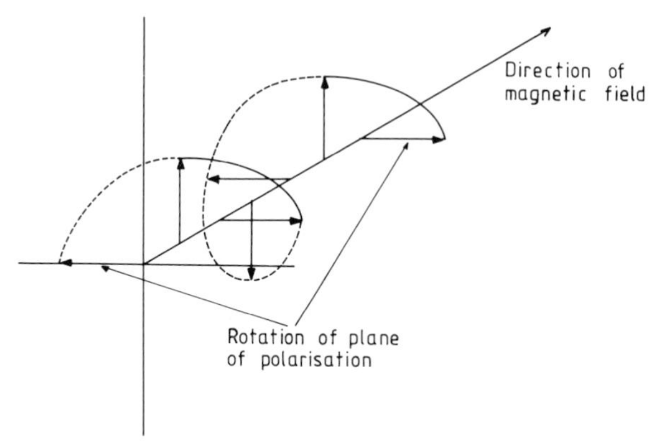

Figure 2.19 Diagram illustrating the Faraday effect for the case in which direction of propagation of the wave is along a line of force.

the presence of an external field the core will become magnetised and this will affect the frequencies induced in the second coil. An analysis of these frequencies can give information on the strength of the field parallel to the axis of the core. By using three separate devices like this, with the axes of their cores at right angles to each other, it is possible to investigate the strength and direction of an external field.

The other main type of instrument makes use of the Zeeman effect. One version of this device is known as the alkali-vapour magnetometer. It uses the fact that if circularly polarised light is shone through vapour of an alkali metal (e.g. rubidium) which is being excited by a rapidly varying magnetic field, the light will be absorbed at certain frequencies which are related to the energy levels in the atoms, and to the strength of any external steady magnetic field.

Summary

In this chapter the essential features of gravity and nuclear forces have been briefly described, and the relevance of these forces to astronomical situations pointed out. The nature of and intimate relationship between magnetism, electricity and electromagnetic radiation has been discussed in more detail. It is this interrelationship which makes it possible to measure magnetic fields in inaccessible celestial objects. However, many vital concepts important to the study of cosmic magnetism can be discovered by looking at the most accessible large-scale field, which is of course that of Earth.

Chapter Three

The Magnetic Field of Earth

Most of us have at some time made use of a magnetic compass to find direction. Einstein recalled how, as a child, he was intrigued by the fact that the compass needle, totally enclosed, insulated and unreachable, could be caught in the grip of an unseen force that made it strive determinedly towards the north. For hundreds of years sailors have used the compass to find their way across the sea. This chapter describes the magnetic field of Earth (called the geomagnetic field) and looks at those properties of the Earth relevant to understanding the effect of the geomagnetic field on various phenomena including different life forms. A history of the Earth's magnetic field is given, together with the theories that could explain its origin.

Brief History of Geomagnetism

Although the Ancient Greeks were aware that a substance called lodestone could attract certain other substances, they do not appear to have discovered the direction-finding properties of magnets. This discovery was first made by the Chinese, although they were not the first to use it for navigation. Chinese divination made use of a variety of different techniques, including their own approach to astrology, and the methods of geomancy—which depended on the influences of the Earth. The main idea in geomancy was that houses of the living and tombs of the dead must be properly aligned if living people wished to enjoy health, wealth and happiness, and avoid evil influences. The shape of the land and the direction of streams, winds and waters were all important to the geomantic diviners. They developed the geomantic compass (figure 3.1) to help them in their craft.

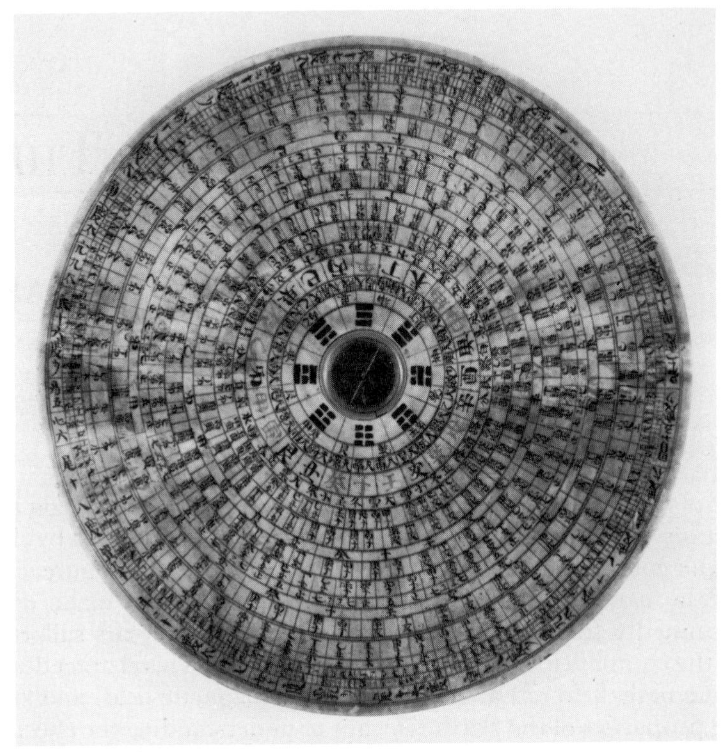

Figure 3.1 A geomantic compass. (Courtesy of the Trustees of the National Maritime Museum, Greenwich.)

The mariner's compass was a development of this device, used for different purposes. Geomancy required a knowledge of the compass directions at different places on the surface of the Earth, and how these directions varied with time. Professor Joseph Needham has described geomancy as the true origin of the science of geomagnetism, much as astrology acted as a stimulus to astronomy, and alchemy foreshadowed chemistry. By the twelfth century knowledge of the compass had reached Europe: Alexander Neckam, an English monk, referred to the compass in a work of that period. There is plenty of historical evidence to suggest that the use of the magnetic compass for navigation was widespread in Europe and Asia by the thirteenth century.

As the ordinary compass needle is only able to move freely in a plane parallel to the Earth's surface it cannot give us all the information we need to plot the magnetic lines of force of the Earth. In order to get more information we also need a magnetic needle that is able to move freely in a vertical plane. Such a device is called a dip circle and one of its earliest recorded uses was by Robert Norman in 1576, so it is often referred to as the Norman dip circle (figure 3.2). The angle between the horizontal and the needle of the dip circle, when the plane of its motion is parallel to the horizontal magnetic field at that point, is called the angle of dip, or inclination. The angle between the direction of true north at a given point and the direction of the compass needle is called the magnetic declination (figure 3.3). The angle of inclination and the declination both vary over the surface of the Earth.

By making use of information on compass directions collected from the worldwide navigation of his day, William Gilbert, physician to Queen Elizabeth I, formulated a theory of the Earth's magnetic field. He was able to show that the magnetic field of a magnetised sphere was very similar to that of the Earth. In other words he showed that the Earth behaved as if it had a bar magnet situated close to its centre, and

Figure 3.2 A dip circle. (Courtesy of the Trustees of the National Maritime Museum, Greenwich.)

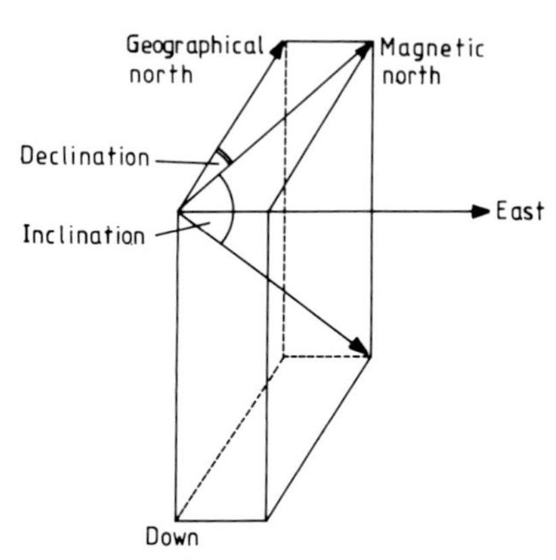

Figure 3.3 The components of the geomagnetic field.

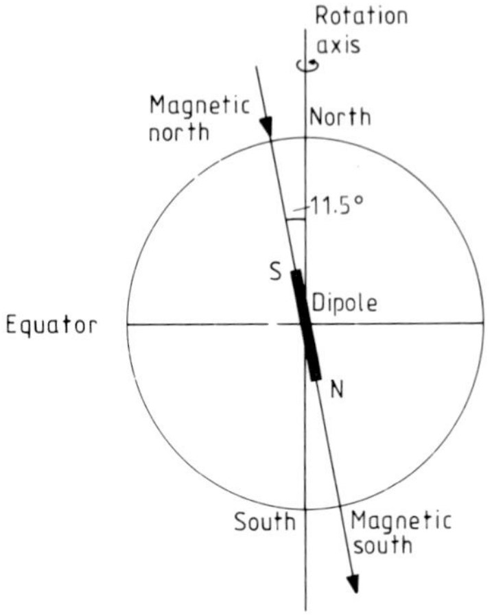

Figure 3.4 The terrestrial dipole.

almost aligned with its rotation axis. Modern work on the magnetic field seems to show that the major part of the Earth's field can be described by imagining a bar magnet situated about 400 km from the Earth's centre, with its axis inclined by 11.5° to the north–south axis (see figure 3.4).

The Magnetosphere

Over the last two decades the use of magnetic measuring instruments on board Earth satellites has greatly increased our knowledge of Earth's magnetic field far above the surface. These satellite measurements have shown that the field is contained within a region—called the magnetosphere—which is compressed on the sunward side and drawn out into a long tail on the opposite side (figure 3.5). The Sun emits a continuous stream of very energetic charged particles known as the solar wind. These particles consist mainly of electrons and protons with traces of other ions and the nuclei of atoms, and they are quite often referred to as solar cosmic rays. As they are electrically charged they cannot cross the lines of force of the Earth's field, but stream past it making a bow shock wave very similar in structure to the bow wave of a ship as it passes through water. Somewhere behind the Earth the various strands of the solar wind join up again, thus enclosing the Earth's field within the pear-shaped region called the magnetosphere. The region behind the Earth is called the magnetotail. Also trapped in the magnetosphere are the charged particles of the Van Allen radiation belts. Near the magnetic poles of Earth the magnetic field becomes

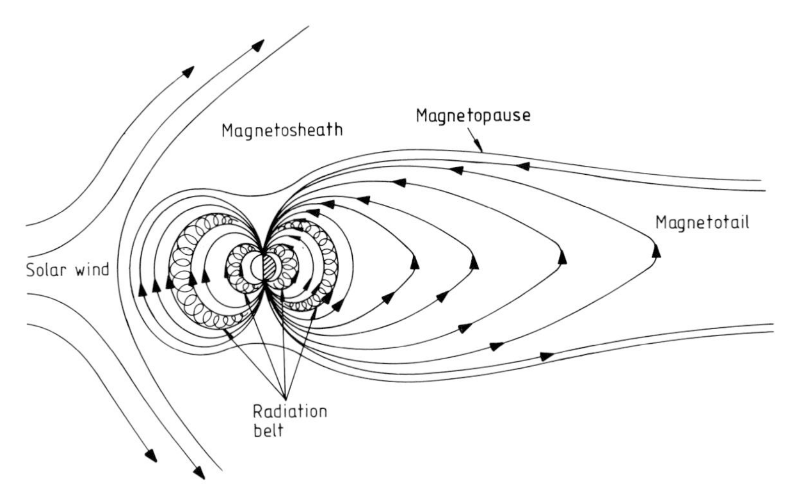

Figure 3.5 The magnetosphere of the Earth.

much stronger and the particles get 'reflected' at these points, and hence they tend to bounce back and forth between the poles and along the field lines. However, as a particle will follow a tighter curve in a strong field than it does in a weak field, and as the magnetic field gets weaker with distance from the Earth's surface, the path of a particle projected onto the plane of the magnetic equator will be as shown in figure 3.6. The combination of the north–south motion and the east–west drift will cause the particle to move within the Van Allen radiation belts, each belt corresponding to a different energy range for the charged particles.

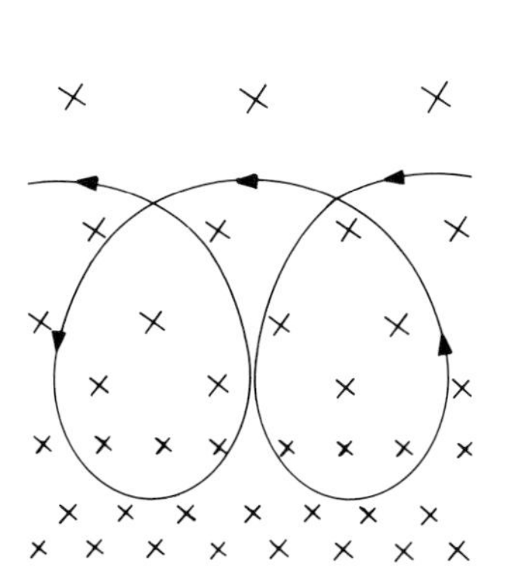

Figure 3.6 The motion of an electron in the plane of the geomagnetic equator.

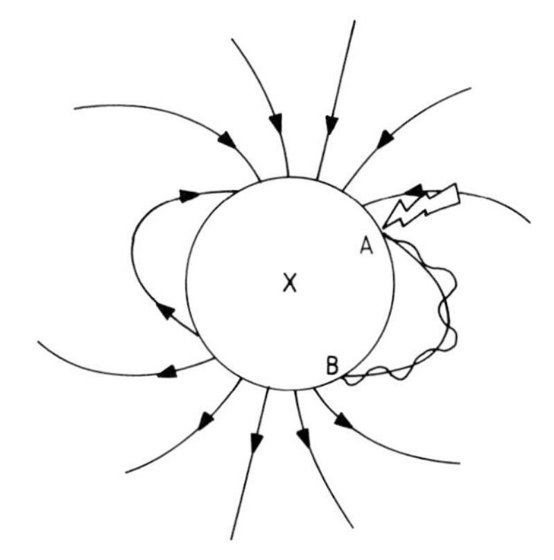

Figure 3.7 The origin of whistlers. Lightning at A causes a disturbance, which travels along a line of force to point B.

Most of us will have experienced a whistling interference on some radio channels, due to phenomena called whistlers. These are caused by low frequency radio noise, generated by lightning discharges, which can travel along the Earth's magnetic lines of force. This means that if there is an electric storm at some place on the Earth's surface, whistlers will affect radio communications at another place where the lines of force originating at the point of the storm enter the interior of Earth (see figure 3.7). There are other interesting phenomena connected with the Earth's field, which are described in the rest of this chapter.

Solar–Terrestrial Relationships

The Royal Observatory at Greenwich was established to provide astronomical data for ship's navigators. Since the position and motion of the Sun are important to navigation it would have been natural for the detailed study of the physical nature of the Sun to develop out of the other work of the Observatory. However, at Greenwich the reasons for setting up a Solar Department came from a rather different direction. In 1843 a German astronomer called Schwabe discovered that the number of dark patches on the Sun's surface—called sunspots—increased to a maximum and then decreased to a minimum with a period of about eleven years. This phenomenon is described in more detail in a later chapter. Historically, it was the discovery in 1852 by a British scientist, Sabine, that the sunspot cycle was related to variations in the compass needle, which led to the establishing of a Solar Department at Greenwich. It was argued that as the main purpose of the Observatory was to serve the interests of mariners, and as Greenwich had a magnetic observatory for this purpose, it was necessary to investigate the links between events on the Sun and the terrestrial magnetic field.

Aurorae and Magnetospheric Substorms

The northern lights, or aurora borealis, has for centuries evoked awe and wonder in people living in high northern latitudes. It is not surprising then that descriptions of this phenomenon are to be found in the poetry, literature and folklore of many northern countries, and that it forms an important part of the culture of the Scandinavian countries in particular. Descriptions of the southern lights, or aurora australis, were first reported by Captain Cook in 1773. Generally these phenomena are called the aurorae polaris, and it has been known since 1722 that these occurrences are associated with changes in the Earth's magnetic field.

The aurorae can best be described as moving curtains of light. Mostly these curtains are greenish-blue, but red aurorae are also seen on some occasions. The extensive use of all-sky cameras, fast jet planes and, more recently, satellite observations, has helped geophysicists to determine the region in which the maximum number of aurorae occur. The region is called the auroral oval and it is actually the intersection of the outer shell of the Van Allen belts with the Earth's atmosphere (see figure 3.8(*a*)). The auroral light is emitted by atoms and molecules of different gases, in the upper atmosphere, which have been excited by collisions with the energetic particles present in this belt.

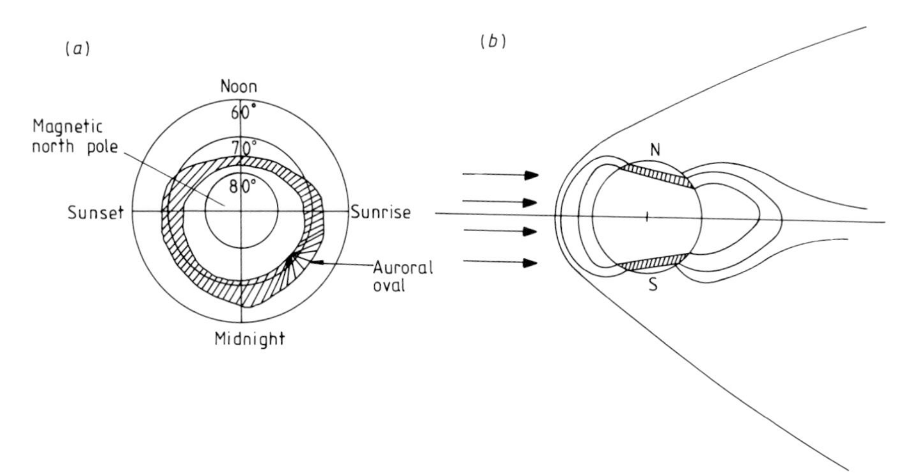

Figure 3.8 (*a*) The auroral oval. (*b*) The distortion of the auroral oval by the solar wind.

There is a very close relationship between the number of electrons of a given energy, which enter the auroral regions of the atmosphere, and the intensity of the auroral light. The energies of electrons are usually measured in electron volts (eV) or kilo-electron volts (keV). An electron volt is the energy gained by an electron as it moves through a potential difference of 1 volt and a kilo-electron volt is 1000 times this basic unit. The close relationship between the number of 6 keV electrons and the intensity of the auroral emission at a wavelength of 4278 angström units (one angström unit is equal to 10^{-10} metre) is shown in figure 3.9. The frequency of aurorae increases with the number of sunspots, and the annual occurrence of red northern lights is also related to the solar cycle (see figures 3.10 and 3.11).

The interaction of the solar wind with the magnetic field of Earth distorts the magnetosphere and consequently it also distorts the auroral oval. This distortion accounts for the fact that the auroral oval is seen at higher altitudes on the dayside and lower latitudes on the nightside of Earth (see figure 3.8(*b*)). With increasing solar activity the auroral oval becomes wider and its lower boundary moves noticeably closer to the equator. The occurrence of aurorae and the width of the auroral oval are related to events on the Sun via phenomena known as magnetospheric substorms.

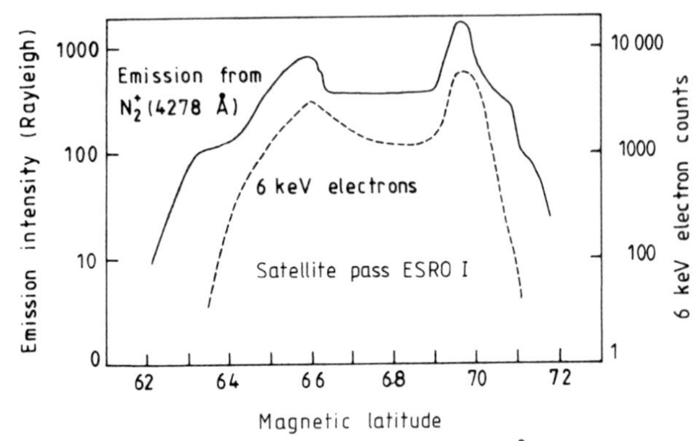

Figure 3.9 The auroral emission at 4.278 Å and the energetic electron flux at 6 keV are similarly distributed in time and space, indicating a very strong relationship. (Figures 3.9, 3.10 and 3.11 from *The Northern Light* by A Brekke and A Egeland (1983) (Berlin: Springer).)

A substorm is the process whereby some of the increase of energy in the solar wind, and changes in the interplanetary magnetic field, are fed into an aurora. Details of the interplanetary magnetic field will be given in a later chapter. Here, let us just see it as strands of the solar magnetic field which have become embedded in the solar wind. To some extent it seems as if the onset of magnetospheric storms is triggered by changes in the direction of these interplanetary strands. However, the first stage in the energy conversion process is accomplished by the interaction of the solar wind with the magnetosphere via a mechanism known as a magnetohydrodynamic dynamo.

Any situation in which an electrically conducting material moves across magnetic lines of force constitutes a dynamo. Figure 3.12 shows schematically the magnetospheric dynamo caused by the flow of the solar wind across the open field lines formed by some geomagnetic field lines which have merged with interplanetary field lines. It is the north–south component of the interplanetary field which controls the production of the open field lines and hence plays an important part in the efficiency of the dynamo.

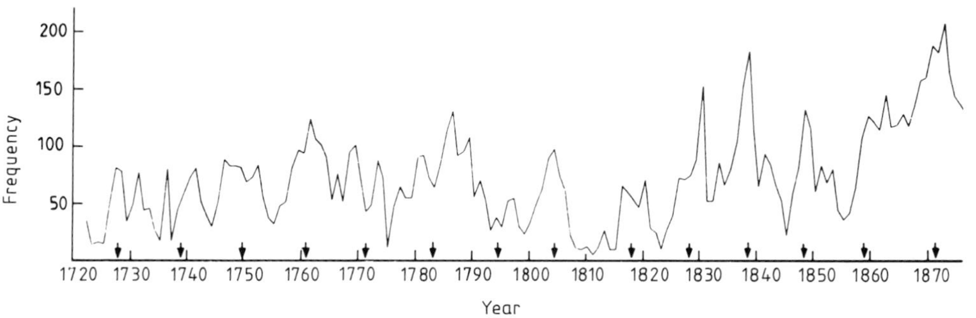

Figure 3.10 Frequency of auroral and sunspot maxima (indicated by arrows).

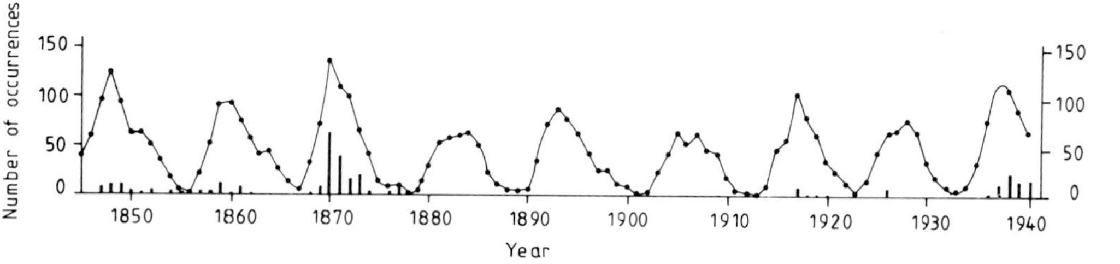

Figure 3.11 The annual occurrence of red northern lights over Norway from 1850–1940 (shown as dark streaks) compared with the annual sunspot curve.

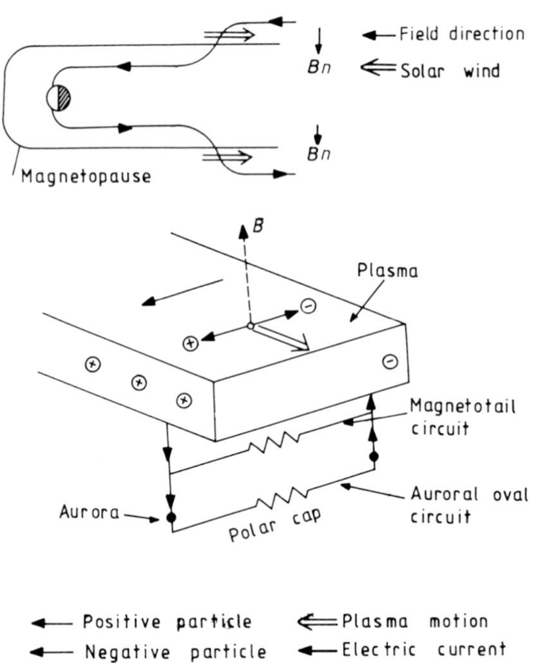

Figure 3.12 Schematic diagram showing the dynamo process associated with the solar wind–magnetosphere interaction (after Akasofu).

One possible way in which the interplanetary magnetic field can influence the magnetosphere is through a phenomenon known as circularly polarised Alfven waves, propagating along the strands of the interplanetary field. Since field lines are somewhat like stretched elastic bands they can be set vibrating just like strings. However, it is also possible to have wave-like disturbances in the form of corkscrews, and this type of wave is called a circularly polarised Alfven wave. In such a wave the magnetic field at right angles to the field lines will be rotating as it reaches Earth. Sometimes it will point in a north–south direction, at other times it will point east–west, then south–north and finally west–east before the cycle starts again. It is the north–south component of the field that has a very marked effect on the dynamo. When this field has a large northward component the efficiency of the dynamo is at a minimum, and its efficiency is increased considerably when the field has a large southward component.

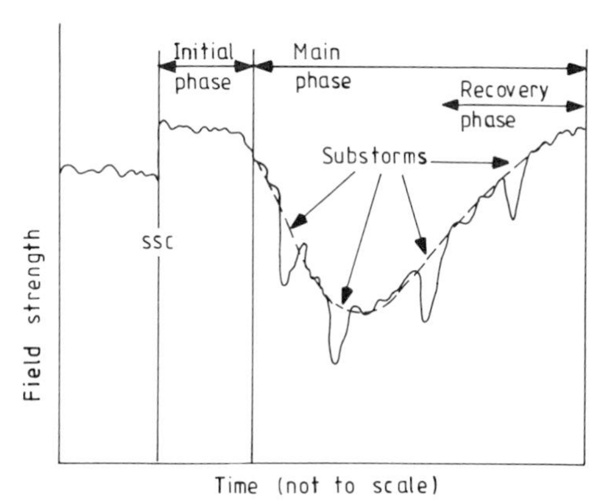

Figure 3.13 Idealised storm variation of the geomagnetic field. The initial phase is attributed to the impact of an enhanced solar wind: the main phase is attributed to an enhanced ring current (after Radcliffe).

The substorms are part of longer-lasting and more extensive changes in the magnetosphere, and consequent variations in the field strength observed at ground level. This is called a magnetic storm, and the behaviour of the northward component of the horizontal field measured at ground level, for an 'average' storm, is shown in figure 3.13. The 'storm sudden commencement' (SSC) is marked by a sudden increase in this component of the field. This increase in field is retained for a few hours, after which there is a more gradual decrease to a magnitude below that experienced during the 'quiet times'. The decrease is the main phase of the storm and it lasts for one or two days. Thereafter there is a gradual recovery to the 'quiet' magnitude. One or more substorms occur during the main phase. The various changes which take place in the magnetosphere during the course of a storm are shown in figure 3.14. The current ring associated with substorms seems at times to flow only in a limited arc of a circle in the magnetosphere, and then to complete its circuit in another arc in the ionosphere. It must necessarily flow up and down field lines to connect these two arcs (see figure 3.15). The two current arcs are called equatorial and polar electrojets respectively.

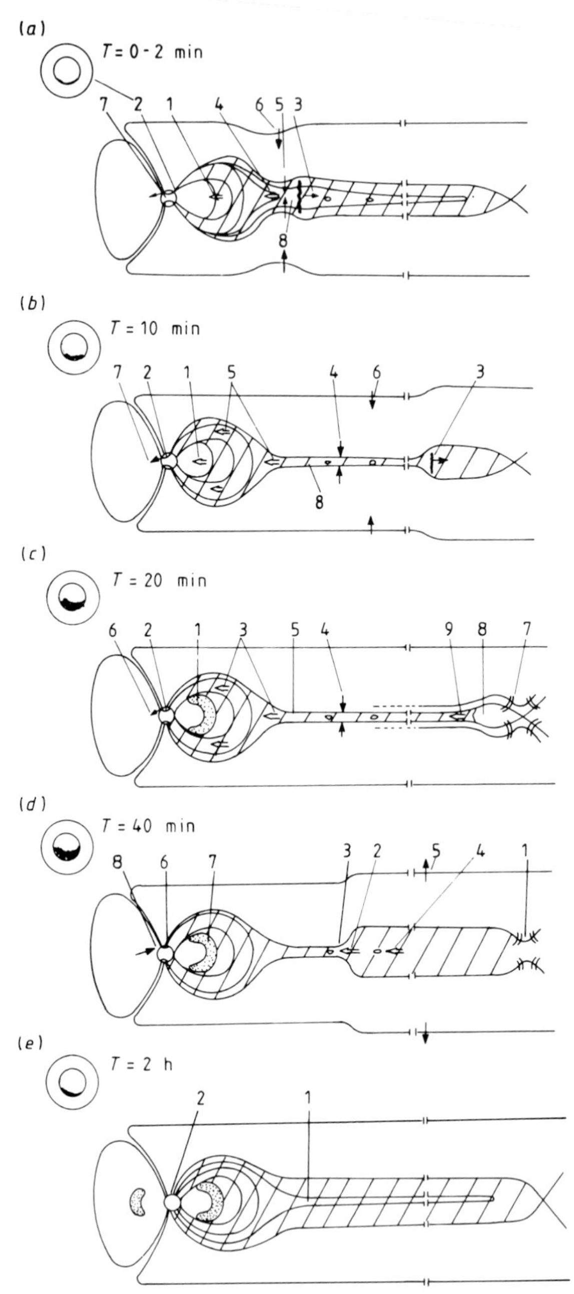

Figure 3.14 Schematic diagrams (noon–midnight cross section) showing various changes which take place in the magnetosphere during the course of a magnetospheric substorm (after Akasofu). *(continued opposite)*

Figure 3.14 (*continued*)

(*a*) $T = 0$–2 min. (1) Sudden Earthward displacement of plasma in the near-Earth plasma sheet (injection). (2) Sudden brightening of an auroral arc and the electrojet formation. (3) Generation and propagation of a fast rarefaction wave. (4) Earthward plasma flow. (5) Plasma sheet thinning. (6) Magnetopause motion. (7) Equatorward motion of the cusp. (8) Disruption and diversion of the cross-tail current.

(*b*) $T = 10$ min. (1) Earthward displacement of plasma. (2) Expanding auroral bulge. (3) Propagation of the fast rarefaction wave. (4) Plasma sheet thinning. (5) Earthward plasma flow. (6) Magnetopause motion. (7) Equatorward motion of the cusp. (8) Disruption and diversion of the cross-tail cusp.

(*c*) $T = 20$ min. (1) Ring current formation. (2) Expanding auroral bulge. (3) Plasma flow. (4) Plasma sheet thinning. (5) Disruption and diversion of the cross-tail current. (6) Equatorward shift of the cusp. (7) Enhanced reconnection. (8) Production of hot plasma. (9) A high speed plasma flow.

(*d*) $T = 40$ min. (1) Enhanced reconnection. (2) High speed plasma flow. (3) Recovery of the plasma sheet. (4) Increase of B_z component. (5) Magnetopause motion. (6) Expanding bulge reaching the highest latitude. (7) Formation of the ring current.

(*e*) $T = 2$ hours. (1) Recovery of the plasma sheet. (2) Contraction of the expanding bulge.

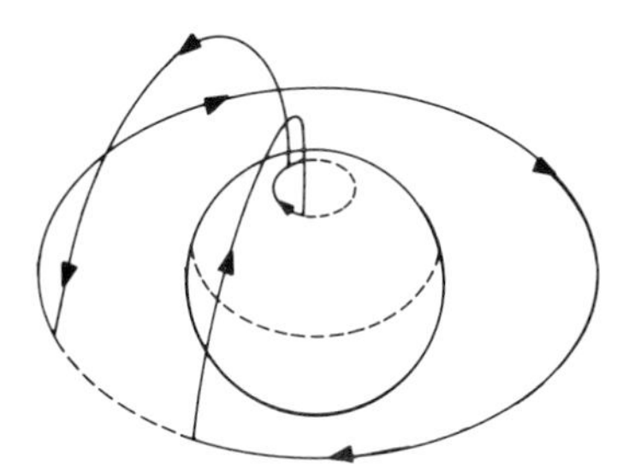

Figure 3.15 A current system that is partly a ring current in the magnetosphere and partly a 'polar electrojet' in the ionosphere (after Radcliffe).

Other Short-term Changes in the Geomagnetic Field

When careful measurements of the geomagnetic field are made over a period of time from any one place on Earth it is clear that the variations of this field are many and varied. These variations are of great complexity, and their study and analysis are a major concern of many geophysicists specialising in geomagnetism. The changes in the magnetic field can be represented by a spectrum which consists of a continuous spectrum of background noise on which are superimposed certain more or less sharp lines. The daily period and its harmonics are the most prominent of these lines (figure 3.16). There is also evidence for lines consisting of the lunar day and its harmonics, but these are less obvious in the figure because they are rather weak. There is also a broad line near 27 days, which may be due to some effect connected with the lunar month or it may be due to effects connected with the rotation of the Sun on its axis. There is also an annual line and one at six months. The causes of some of these lines are not fully understood at the moment. The mean annual magnetic variations for highly disturbed, moderately disturbed and weakly disturbed years are shown in figure 3.17. The daily variations in the north–south, east–west and vertical components of the field are latitude dependent. This dependence is shown in figure 3.18.

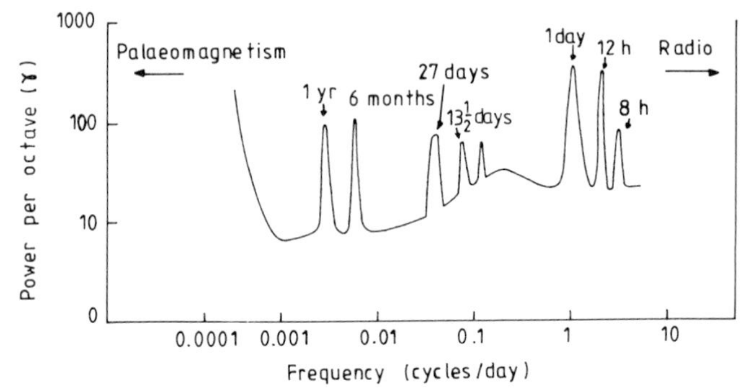

Figure 3.16 The power spectrum of the horizontal component of the magnetic field at Greenwich, Abinger and Hartland. The spectrum has been prepared by adjusting and joining together sections derived from observations taken at different periods. It gives a general idea of the form of the curve over a wide range of frequencies but is to be regarded only as a generalised diagram (after Bullard).

Some of these variations of the geomagnetic field can affect a few ground-based technological systems, especially in or near the auroral oval and during magnetospheric substorms. Not long after the installation of telegraph and telephone systems it became clear that serious disturbances could occur during those periods when the aurorae were most active. One investigator of auroral phenomena, Professor Carl Størmer, made use of currents induced in telephone wires to activate an alarm to indicate the presence of the northern light. Telephone companies have designed and installed equipment which protects telephone networks against the possible damaging effects of these currents. Measurable currents are also induced in power lines during severe magnetic disturbances, and there have been reports from America of transformer breakdowns during high solar active periods. Currents induced in the transatlantic cable between Newfoundland and Scotland, during a severe storm in 1940, led to distortion of

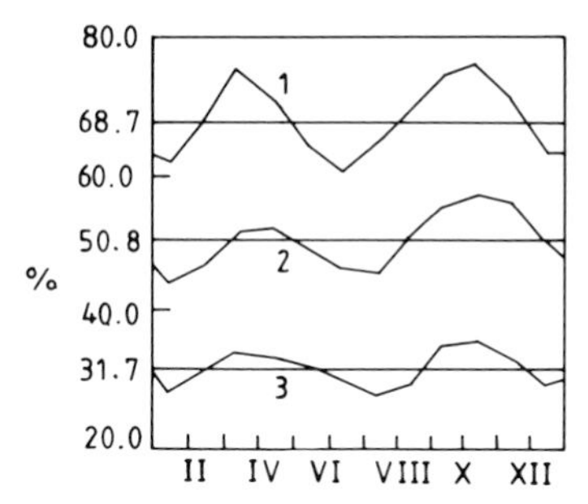

Figure 3.17 The straight lines at 1, 2 and 3 represent the mean annual magnetic activity (%) for (1) a highly disturbed year, (2) a moderately disturbed year and (3) a weakly disturbed year. The curves represent the annual deviations about the mean. The figures on the bottom axis are months of the year. (From *The Geomagnetic Field and Life* by A P Dubrov (1978) (New York: Plenum).)

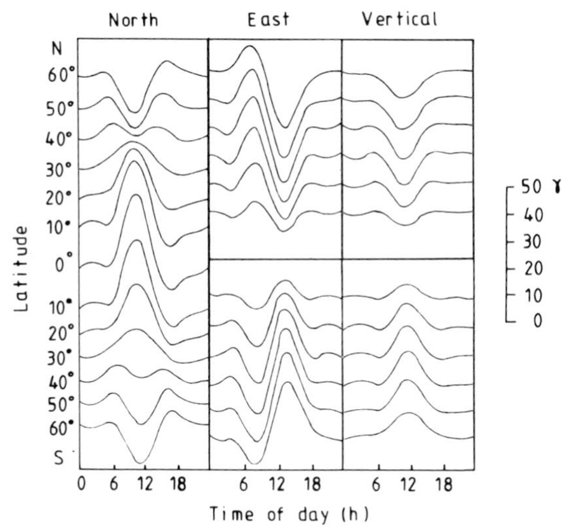

Figure 3.18 The form of the daily variation as a function of latitude at the equinoxes of 1902, a year of sunspot maximum. (From *Geomagnetism* vol 1 by S Chapman and J Bartels (1940) (Oxford: Oxford University Press).)

conversations in one direction though not in the other. It has been suggested that some shipwrecks and aircraft crashes which occurred during intense auroral displays may have been due to the deviation of the compass needle. Auroral displays are also known to disrupt radio communications in and near the polar regions.

Besides the short-term variations in the Earth's field, there are also longer-term variations covering hundreds, thousands and even millions of years.

History of Earth's Magnetic Field

There are three methods used to investigate the history of the Earth's field. Actual measurements of the strength, dip and declination covering a long period of time are the most direct method of investigating long period changes. However, this is of course limited to the period for which accurate observations have been recorded, and systematic observations which can be used for this purpose really only started in 1635. Accurate observations cover an even shorter period—from 1830 to the present day. These observations show that not only has there been a change in the strength of the field, but the positions of the geomagnetic poles have also changed (see figures 3.19 and 3.20).

The next method makes use of archeological data. Many civilisations in various parts of the world had kilns which had been baked to very high temperatures, and tiny magnetic fragments in the floors and walls of these kilns would have been heated above the Curie temperature. This would mean, as explained before, that any magnetism present in the fragments would have been destroyed. As the kilns cooled the tiny

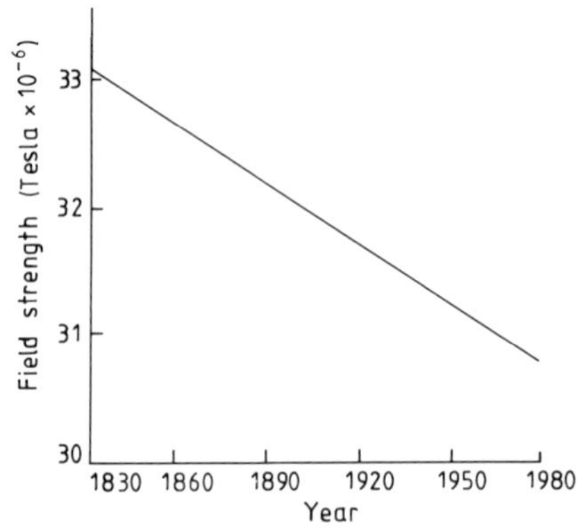

Figure 3.19 Variation of the strength of the geomagnetic field with time. (Courtesy the Open University.)

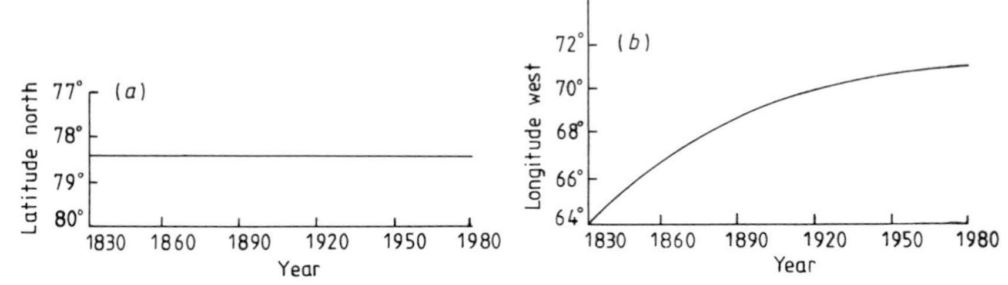

Figure 3.20 (*a*) Variation of latitude of north magnetic pole with time. Almost constant at about 78.5°. (*b*) Variation of longitude of north magnetic pole with time. (Figures 3.20 and 3.21 from *Earth's Magnetic Field* Science Foundation Course Unit 23 (1971) courtesy the Open University Press.)

particles would be re-magnetised in the direction of the Earth's field as it was when the kiln was last fired. The number of magnetic fragments aligned by the field is related to the strength of the Earth's field at that time, so we can obtain information about the strength and direction of the field at various times by studying the magnetic particles in kilns from various ancient civilisations. This method is particularly useful for investigations of the field over the past few thousand years because in many cases the samples may be accurately dated (using other archeological methods). The variation of the strength of the field with time obtained using this method is shown in figure 3.21.

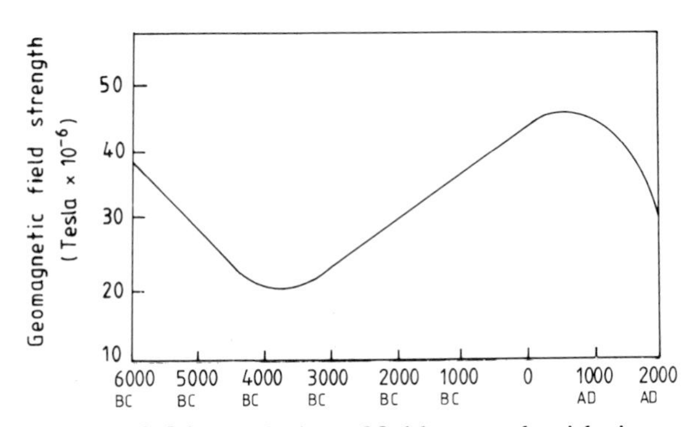

Figure 3.21 Variation of field strength with time.

The third method is based on geological data and provides information on changes in the field covering a period of about five million years. This method is based on the fact that almost all rock contains small quantities of iron compounds and when a rock is formed it becomes magnetised in the direction of the Earth's field as it is at the time of formation. Once again, the strength of the magnetism is related

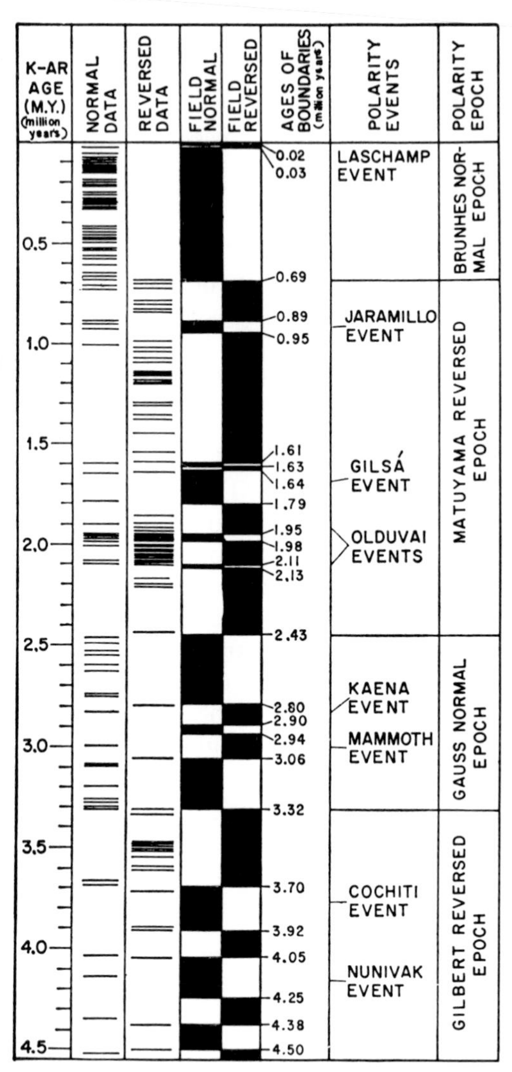

Figure 3.22 Timescale for geomagnetic reversals. Each short horizontal line shows the age as determined by potassium–argon dating and the magnetic polarity (normal or reversed) of one volcanic cooling unit. Normal polarity intervals are shown by the solid portions of the 'field normal' column, and reversed polarity intervals by the solid portions of the 'field reversed' column. The duration of events is based in part on palaeomagnetic data from sediments and profiles. (Reproduced with permission from *Reversals of the Earth's Magnetic Field* by J A Jacobs (1984) (Bristol: Adam Hilger).)

to the strength of the field at that particular time. The age of the rock can be found by geological methods. The measurements made using this method produced the surprising result that the direction of the Earth's field has reversed several times during the last four and a half million years (figure 3.22). These long-term variations of the strength and direction of the field naturally led scientists to ask questions about the origin of the Earth's magnetic field. However, before theories on the origin of the field can be described it is necessary to look briefly at the properties of the Earth.

Relevant Properties of the Earth

The Earth is not an exact sphere, but is slightly flattened at the poles and bulges slightly at the equator (see figure 3.23). When explaining the seasons (figure 3.24) it is generally assumed that the Earth's axis always points in the same direction with respect to the stars, making an angle of 66.5° with the plane of the Earth's orbit. This is true to a high

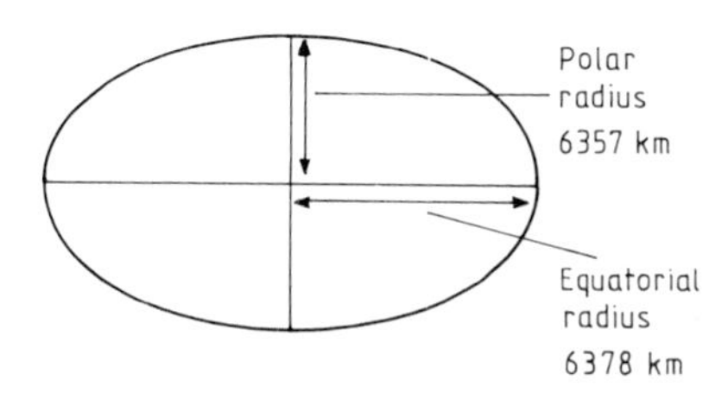

Figure 3.23

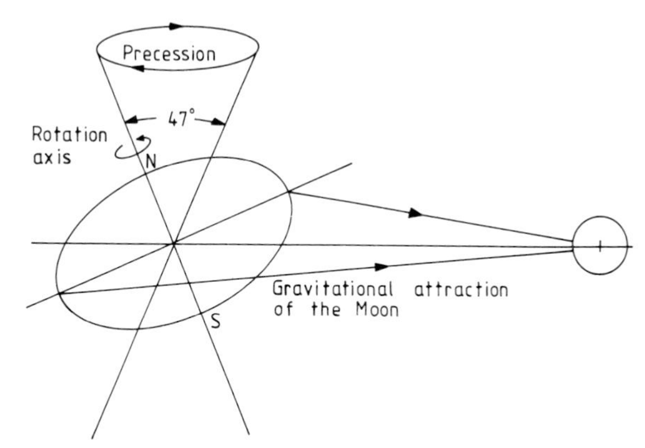

Figure 3.24 Precession of the Earth's axis.

degree of accuracy over short periods of time (that is to say, short compared with 26 000 years). However, the combined pull of the Sun and the Moon on the equatorial bulge of Earth causes Earth's axis to sweep out a cone, in much the same way as the axis of a child's top will seem to move on the surface of a cone as the top slows down. This is called the precession of the Earth's axis, and the time it takes to sweep out one complete cone is 26 000 years. In addition, there is a slight nodding of the axis due to the fact that the gravitational pull of the Sun and the Moon are in directions that are constantly changing. This nodding is known as the nutation of the axis, and it has a period of 18.6 years.

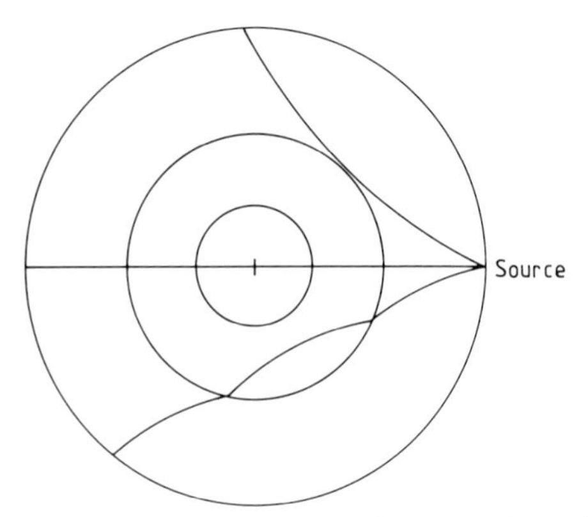

Figure 3.25 The passage of earthquake waves through the Earth.

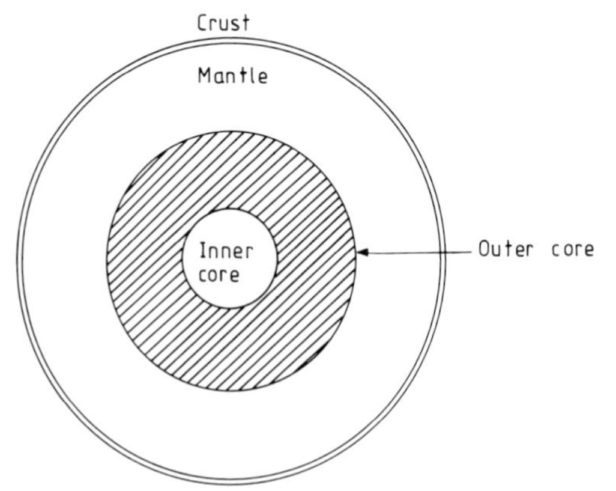

Figure 3.26 Internal structure of Earth.

A great deal of information on the internal structure of Earth is obtained by studying seismic waves, originating during earthquakes, arriving at different geophysical observatories (see figure 3.25). These data have been combined with other data on the average density of the Earth, compared with the density of the surface rock, to construct a picture of the internal structure of Earth (figure 3.26). The outer core probably consists of an iron–nickel alloy in liquid form, and because of its fluid and electrical conducting properties it plays an important part in one theory of geomagnetism.

The Origin of the Geomagnetic Field

The first theory on the origin of the Earth's magnetic field was put forward by William Gilbert in 1600. He showed that a magnetised sphere of lodestone had a magnetic field of the same shape as that part of the Earth, and from this he proposed that the Earth was a uniformly magnetised sphere. The first objection to this idea is that a few tens of kilometres below the Earth's surface the temperature is high enough to destroy any magnetism in iron-bearing minerals, i.e. the temperature is above the Curie temperature of these minerals. In other words the thermal motion of the particles of matter at these depths will be sufficient to destroy the alignment of the domains necessary for permanent magnetism. This means that if the Earth's field is due to permanent magnetism, this must exist in a thin spherical shell of the Earth's crust. However, as the materials of this thin shell are not magnetic enough to explain the observed strength of the field the theory becomes unacceptable. The other reason for rejecting this theory is that it cannot explain the observed changes in the field.

A much more convincing theory is that motion in the outer core can give rise to what is called a self-sustaining dynamo. In Chapter Two it was seen that if a conductor is moving in a magnetic field an electric current can flow in the conductor if it is part of a complete electric circuit. If the Earth had some intrinsic magnetism to start with, then the motions of the electrically conducting outer core would give rise to electric currents. These currents could give rise to a magnetic field and this field could in turn sustain its own current, provided there were a source of energy to sustain the motions in the core. A model of such a dynamo is shown in figure 3.27.

There are three possible mechanisms that could give rise to the motions of the outer core which are necessary for this self-sustaining dynamo to be applicable to Earth. The first is similar to the thermal convection currents seen when, for example, heating soup in an open pan. The soup in contact with the bottom of the pan will rise to the surface since, being hotter, it is less dense than the rest of the soup, and the cooler parts of the soup will sink towards the bottom, setting up convection currents. These currents are maintained by the heat below the pan. In the case of the Earth, atoms of radioactive substances such as uranium, thorium and plutonium are continually breaking up into smaller atoms and subatomic particles. This process is called radioactive decay, and heat is produced as a result of the process. The decay of these substances in the core of the Earth could provide the source of heat necessary to maintain the thermal convection of the Earth's dynamo.

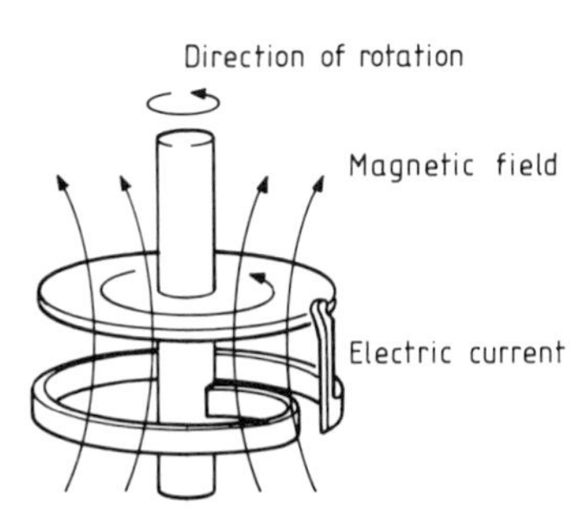

Figure 3.27 A model of a self-sustaining dynamo. (Reproduced with permission from *Reversals of the Earth's Magnetic Field* by J A Jacobs (1984) (Bristol: Adam Hilger).)

We have already seen that Earth's axis precesses due to the gravitational pull of the Sun and Moon on the equatorial bulge of Earth. This provides a second possible source of energy to sustain motion in the outer core. An experiment can demonstrate this. Suppose a large basin is filled with water, and a plastic disc with an arrow on it is floated on the water, and the basin is then rotated gently. The motion of the basin will not immediately be taken up by all parts of the water and this will be evident from the behaviour of the disc. If the arrow is pointing in a given direction—say east—before the basin is rotated, it will continue to point in this direction when the basin is rotated. The plastic disc with the arrow on it will only gradually take up the rotation of the basin. In a similar way, because the outer core of Earth is fluid it does not immediately take up the changes in motion of the solid parts of Earth, and the different rates of precession between core and mantle could give rise to complicated motions that could sustain a dynamo.

The third mechanism is associated with the separating out of dense materials from less dense materials in the outer core, due to the growth of the inner core. The dense material will move downward and the less dense material upward. This movement will mean a change in the gravitational energy of the outer core and this energy will thus be available to generate the magnetic field. Several geophysicists favour this particular mechanism which at present seems the most likely explanation of the source of energy for powering the generation of the Earth's magnetic field.

Biological and Meteorological Effects of the Geomagnetic Field

In recent years it has also become evident that the Earth's field and its changes can influence biological organisms. In 1975 Dr Blakemore, a biologist working in America, discovered that certain bacteria were influenced by the Earth's field in that under a microscope they would be observed to swarm consistently towards magnetic north. This discovery was confirmed by experiments with simple bar magnets. Further experiments showed that the bacteria were moving towards the north because they wanted to move downwards. The same bacteria swim towards the south in the southern hemisphere, once again because they want to move downwards towards the sediment and away from the surface of the water where the concentration of oxygen, which is toxic to them, is higher. When the bacteria were studied under the electron microscope they were seen to contain a string of 'beads', each bead consisting of a magnetic substance known as magnetite.

Human navigators consciously use the Earth's magnetic field whenever they use a magnetic compass. Evidence has been growing over several years to suggest that birds and other animals are also able to use the geomagnetic field as a navigational aid, although they are also able, when clear skies permit, to use the Sun and stars for the same purpose. Bees, on the other hand, seem to use the polarisation of light in their navigation, but they use daily rapid fluctuations of the geomagnetic field to reset their own internal biological clocks. Some scientists in Britain have shown that most human beings have the ability, to a greater or lesser extent, to navigate using the Earth's magnetic field without the aid of a compass. There seems to be some evidence to support the claim that certain species of micro-organism become extinct following a reversal of the Earth's field, although this evidence is not widely accepted. However, several biologists now believe that the geomagnetic field is important to many forms of life.

Another interesting discovery, this time between weather and the geomagnetic field, was made by Goesta Wollin. He showed that changes in sea surface temperatures were somehow linked to changes in the geomagnetic field, and explained it as follows. The water of the oceans is electrically conducting. Changes in the geomagnetic field cause currents to flow in the oceans, or affect currents already flowing. This causes changes in sea surface temperature patterns, which in turn affect the temperature above the ocean. Long period changes in climate have also been linked to long-term changes in the solar cycle. This will be one of the topics discussed in the next chapter.

Concluding Comments

The geomagnetic field is more readily accessible to investigation than any other large-scale cosmic field and therefore has been studied in more detail. This knowledge of the geomagnetic field, the magnetosphere and its interaction with the solar wind, has led to the development of theories which serve as models for investigating the magnetic fields and magnetospheres of the other planets. However, because of the interaction between the terrestrial magnetic environment and activity on the Sun, it is important to know more about solar magnetic fields.

Chapter Four

Solar and Interplanetary Magnetic Fields

This chapter looks at the nature and behaviour of magnetic fields of the Sun and their extension into interplanetary space. Since these fields are intimately bound up with the structure and mechanics of the Sun, the dynamics of its photosphere, chromosphere and the corona, it is necessary to start with a brief review of some relevant facts about our Sun.

The Sun

The Sun is a fairly ordinary star. It is, however, the star about which the planets of the solar system move, and it is the ultimate source of heat, light and energy for life on Earth. Since it is the most dominant extraterrestrial object it is not surprising that many peoples in different parts of the world should have worshipped it as a god, and that it has been a major part of the serious study of astronomy for thousands of years. The points on the horizon at which the Sun rises and sets at different times of the year were the basis for the first social use of astronomy—that of calendar-making—in many ancient cultures. The varying lengths of the shadows cast by the Sun were one of the earliest methods of time-keeping. The Sun was also used by Eratosthenes—keeper of the library at Alexandria—to measure the size of the Earth in about 240 BC. In more recent times, over the last 400 to 500 years, the Sun has proved to be an invaluable aid to navigation.

The study of the detailed physical nature of the Sun had to await the invention of the telescope and the development of spectroscopy. These developments revealed the detailed behaviour of sunspots and the chemical composition of the Sun's atmosphere. Progress in atomic and nuclear physics led to the construction of mathematical models of solar and stellar interiors. According to the currently accepted model of the Sun, its interior consists of three main regions: the core, the radiative zone and the convective zone (figure 4.1). In the core energy is generated by nuclear reactions, largely the conversion of hydrogen to helium. This energy is transported outward from the core by radiation in the radiative zone and by convection in the convective zone. The magnetic field of the Sun is believed to be due to a dynamo action which exists in the convective zone. Here the thermal convection currents of ionised plasma amplify any weak remnant field, due to the collapse of the Sun out of the gases of the interstellar medium, to create a stronger field. These fields play a vital role in solar activity.

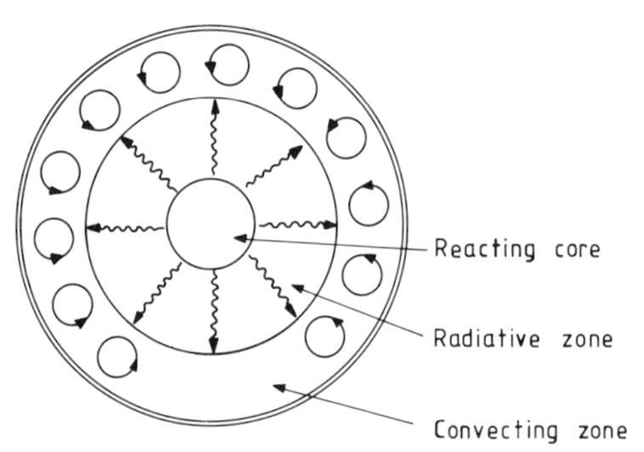

Figure 4.1 Internal structure of the Sun.

The visible surface of the Sun is called the photosphere, and it is the source of the characteristic absorption spectrum of the Sun. Where the photosphere has a boundary with the convective zone, its temperature is about 6000 K. Its temperature drops to 4000 K where it borders with the more tenuous solar atmosphere. The temperature of the chromosphere then increases outwards and reaches 50 000 K, where it meets the outermost part of the solar atmosphere, which is called the corona.

Sunspots

Sunspots were first seen with the aid of telescopes in about 1610, and ever since have been of intense interest to astronomers. Although they appear to be very dark regions of the photosphere, this is entirely a contrast effect. The region of the sunspot is cooler than the surrounding areas of the Sun, so emits less light and as a result looks darker. However, most spots are as bright as the full Moon. When a sunspot approaches the limb of the Sun, the near side becomes practically invisible, whereas the far side is enlarged: this is called the Wilson effect and indicates that sunspots are depressions in the photosphere (figure 4.2). More often than not sunspots occur in pairs or more complex groups (figure 4.3) and whereas the smaller spots, known as pores, last only a few hours, the larger ones last for several days. Observations on the longer-lasting spots show that the Sun is rotating, but not as a solid body. The rotation rate near the poles is about 37 days, whereas it is about 26 days in the equatorial region.

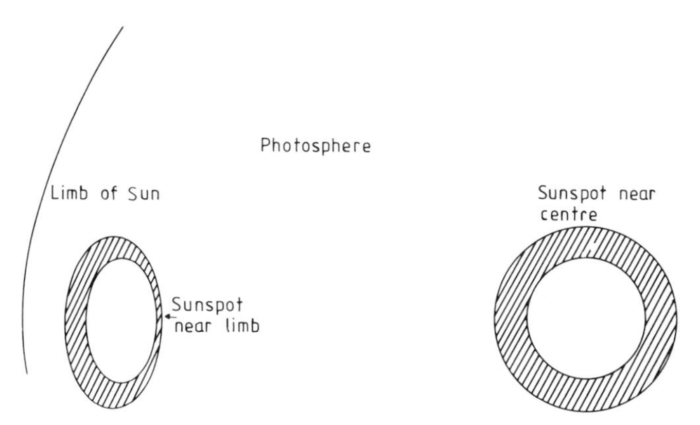

Figure 4.2 The Wilson effect.

The number of spots also varies and reaches a maximum about every eleven years, but the period between successive maxima can be as short as seven or as long as seventeen years. Each cycle begins with the formation of spots in middle latitudes of each hemisphere, about 40–50° from the equator. Subsequently, spots form in lower latitudes until most of the photosphere is covered. After maximum coverage it is the high-latitude spots that disappear first, and the last to fade completely are those close to the Sun's equator. This behaviour is very graphically shown in the Maunder butterfly diagram, which shows the location of sunspot formation versus time in years (see figure 4.4).

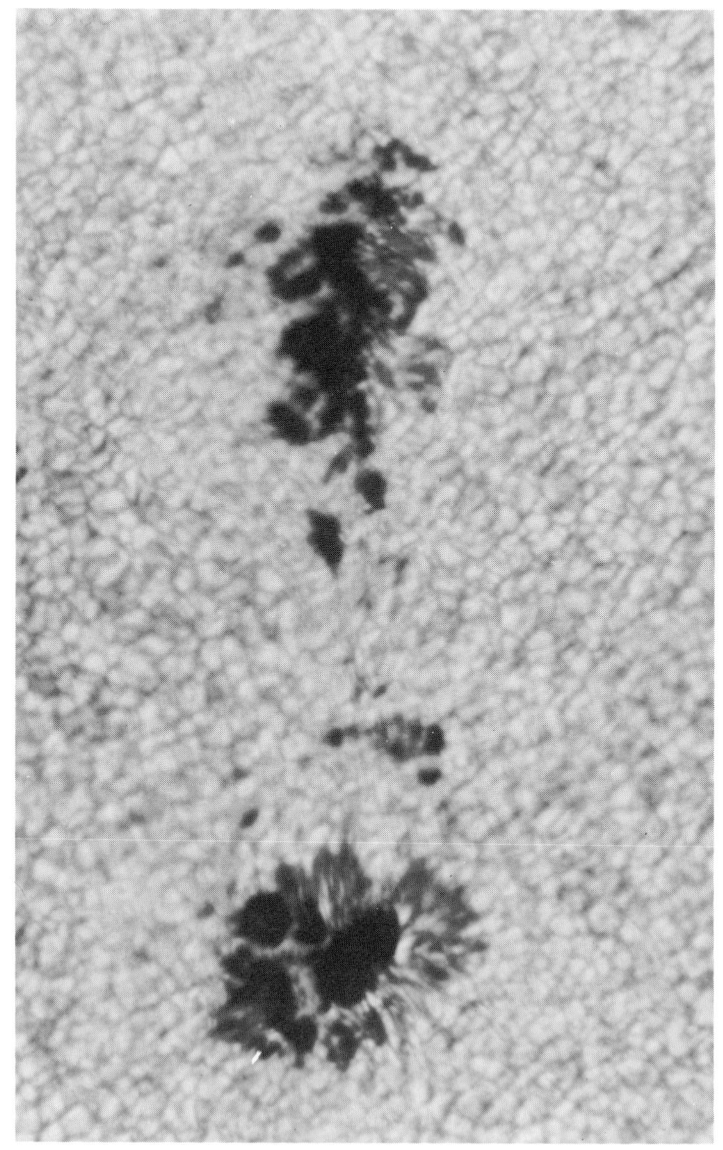

Figure 4.3 Solar granulation and a small growing group of spots and pores. (Courtesy Sacramento Peak Observatory.)

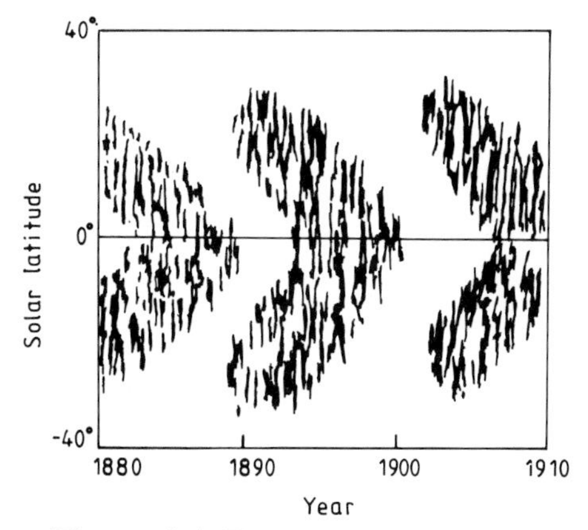

Figure 4.4 The Maunder butterfly.

Other Signs of Solar Activity

Photosphere faculae are bright areas which usually eventually engulf a sunspot group, but are usually noticeable before the appearance of the sunspots, and often last longer than the spots themselves. They are the first signs of solar activity and when they are near the limb of the Sun they appear brighter than the photosphere. Associated with the photospheric faculae are chromospheric faculae or plages, which occur in the chromosphere just above the other faculae. These, however, can only be seen in photographs of the Sun taken with special filters.

A flare is a sudden local increase in the surface brightness of the Sun, occurring in a region active with faculae and sunspots. The effect is produced by the sudden release of tremendous amounts of energy in the upper chromosphere, and is the culmination of the activity that has been building up in the sunspot region.

Solar prominences are another visible consequence of solar activity. When seen at the limb of the Sun (figure 4.5) they appear as luminous arch-like structures with continual internal motion, but when projected on the luminous disc of the Sun they just appear to be dark filaments. Sunspot prominences (also sometimes called active prominences) appear over a sunspot group, whereas quiescent prominences are associated with particular regions without sunspots or with decaying groups of spots.

In 1960 an astronomer called Leighton used the Doppler effect to investigate motions in the Sun's photosphere. He discovered that the Sun seems to be divided into about 1000 larger cells by the horizontal velocity of the photospheric gas which changes in speed and direction from one cell to the next. These regions are known as supergranules to

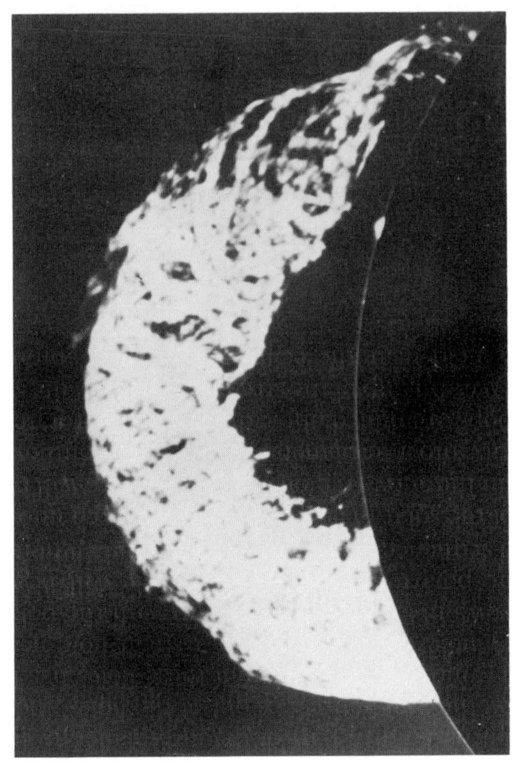

Figure 4.5 Photograph of a solar prominence. (Courtesy of the Royal Astronomical Society.)

distinguish them from the well known granular pattern. At the boundaries of these supergranules there is a concentration of many small jets, known as spicules, which shoot out through the chromosphere into the corona (figure 4.6).

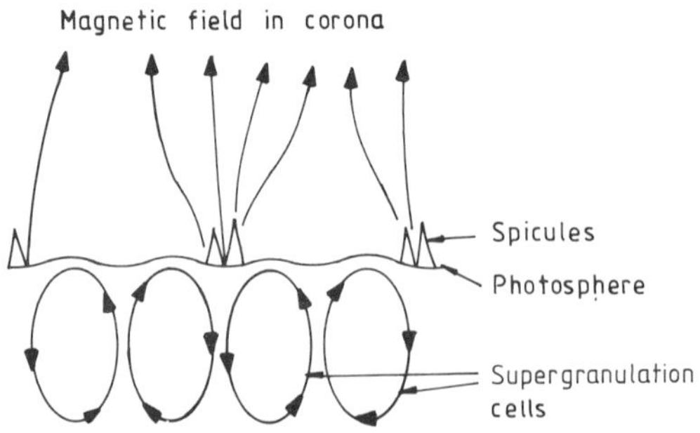

Figure 4.6 Spicules, supergranulation and the solar magnetic field.

Having reviewed the general features of solar activity, we can now look at observations on the magnetic fields of the Sun and relate these field structures to the dynamics of the features described above.

Solar Magnetic Fields

The Zeeman effect (described in Chapter Two) was first applied to astronomical objects by George Hale at Mount Wilson Observatory, when he used it to measure the field strength of sunspots. He was able to show that when sunspots occurred in pairs the magnetic field emerged from below the Sun's surface at one spot and entered the surface again at the other spot. Moreover, during one eleven-year sunspot maximum the spots in the two hemispheres had different east–west polarities from each other, but during the next cycle the polarities in each hemisphere would be reversed (figure 4.7).

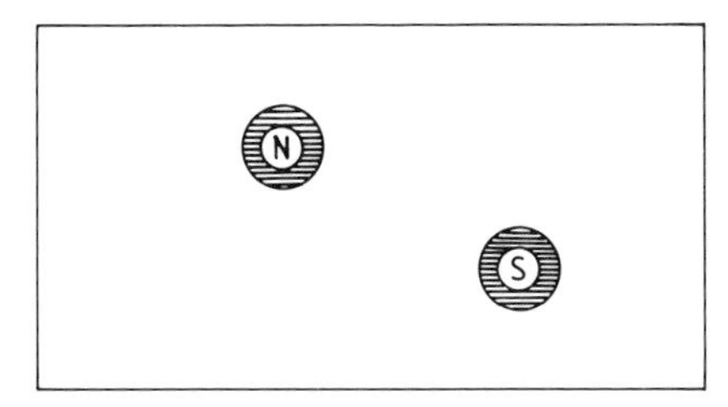

Figure 4.7 Reversal of magnetic polarities in a sunspot pair.

Later work showed that the Sun has fields of opposite polarity at its two poles, and this polarity changes at the start of a new sunspot cycle. There is also a weak large-scale field which, like the spicules, seems to concentrate at the boundaries of the supergranular cells, in some quiet areas of the Sun. Magnetic fields are also found in prominences seen against the disc, and quite often these occur between the magnetic areas of opposite polarity that are associated with sunspot pairs. Astronomers have used special mathematical techniques to analyse the data on the magnetic field distribution over the entire surface of the Sun and from this have calculated the field in the corona. An example of the results from such calculations is shown in figure 4.8.

Besides radiating light and radio waves, the Sun also emits some x-rays—the same type of radiation as used in medical diagnosis. Skylab took x-ray photographs of the Sun which revealed the appearance of 'holes' in the solar corona. The magnetic field data were analysed for the days on which the photographs were taken, and when the results were compared with the photographs they showed that the magnetic field lines seemed to diverge from these coronal holes (figure 4.9).

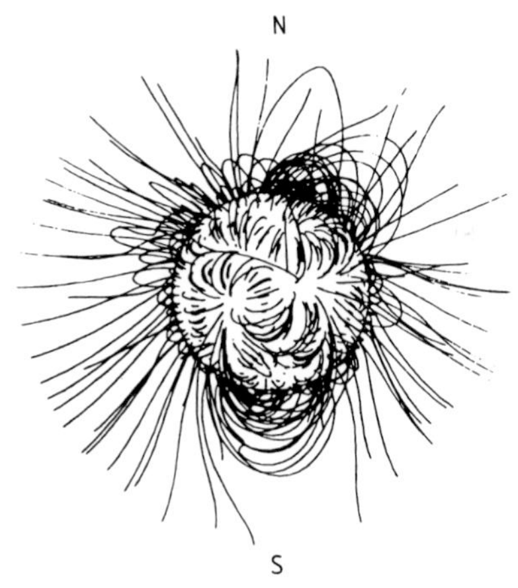

Figure 4.8 Magnetic fields in the solar corona. (Reproduced with permission from Large-scale solar magnetic fields by R Howard *Ann. Rev. Astron. Astrophys.* **15** © 1977 Annual Reviews Inc.)

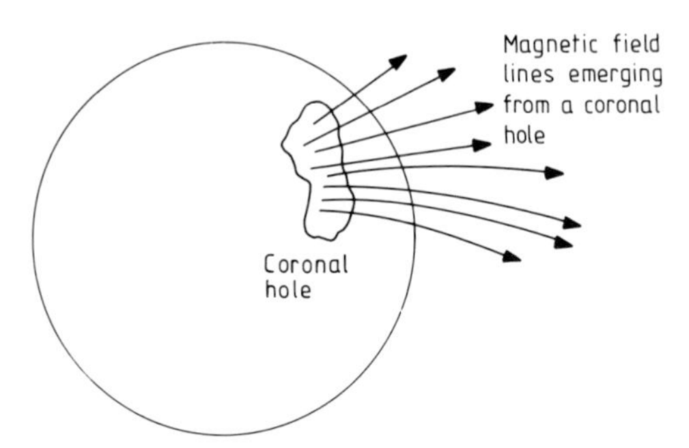

Figure 4.9 Magnetic field lines diverging from a coronal hole.

Magnetic Fields and Solar Activity

From the observations described so far it is obvious that the magnetic fields of the Sun are very closely related to solar activity. Before describing a theory that can account for this relationship let us consider the behaviour of magnetic fields in very hot ionised gases. In Chapter Two we saw that fields associated with a particular magnet would be fixed. However, the field lines could change their shape and reconnect with others if several magnets were moved with respect to each other.

We also saw that the magnetic field of a coil carrying a current was in fact fixed with respect to the coil, and when the coil was moved the field lines would move with it. These ideas can be used to explain the interaction between hot gases, called plasmas, and the magnetic 'lines of force' which thread their way through such plasmas.

Since the atoms of a plasma are ionised it will consist mainly of positively charged ions and free electrons. It is the relative motions of these particles that can generate a magnetic field. If part of the plasma were moved across the field lines a current would be induced into this part and this current would give rise to an additional magnetic field. The original field plus the new induced field will have field lines which look as if the original field lines have just been displaced by the same amount as the part of the plasma that was shifted (figure 4.10). Physicists describe this phenomenon by saying that the magnetic field lines are frozen into the plasma. In an earlier chapter the field lines were described as being like elastic bands in that they have a tendency to contract along their lengths. Hence the field lines will oppose the motion of a small volume of gas across them, and any attempt at such motion would meet with resistance like pulling on a taut elastic band. The field lines pointing in the same direction will tend to repel each other, thus giving rise to a magnetic pressure at right angles to the direction of the lines. This will be in addition to the normal gas pressure within the plasma. These ideas may now be applied to a description of the magnetic behaviour of the Sun.

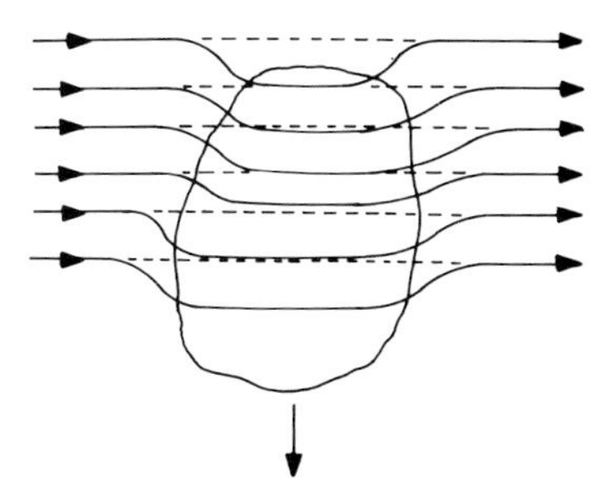

Figure 4.10 Magnetic lines of force frozen into a plasma.

Suppose there is a single strand of a magnetic force line stretching from the north to the south pole of the Sun, immediately below the Sun's surface. Because the equator of the Sun is rotating faster than the polar regions this line will soon be distorted, and after a few rotations it will be wound up as shown in figure 4.11. Because the line is stretched it will have more energy stored in it, just as energy can be stored in an elastic band. In reality there will be several strands of magnetic field

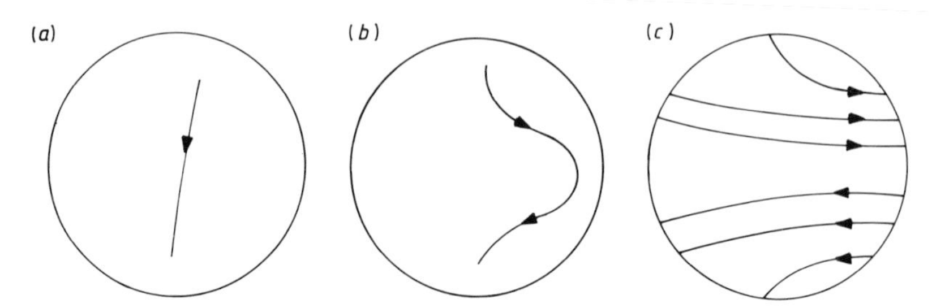

Figure 4.11 (*a*), (*b*), (*c*) The winding up of the solar magnetic field due to differential rotation of the Sun.

lines stretched taut below the surface of the Sun, and since these field lines are in the same direction they will repel each other, thereby creating a magnetic 'bubble' in which the gas density is less than the surroundings. As a result this elongated bubble will rise and eventually break through the surface of the photosphere. The momentum of the bubble will take it high above the surface, and as the ends of the lines are still embedded in the Sun there will be a magnetic tube of force forming an arch above the surface. This is a loop prominence and the two points at which the lines actually cross the photosphere will be cooler than the surroundings. It is these regions that we see as a sunspot pair (figure 4.12).

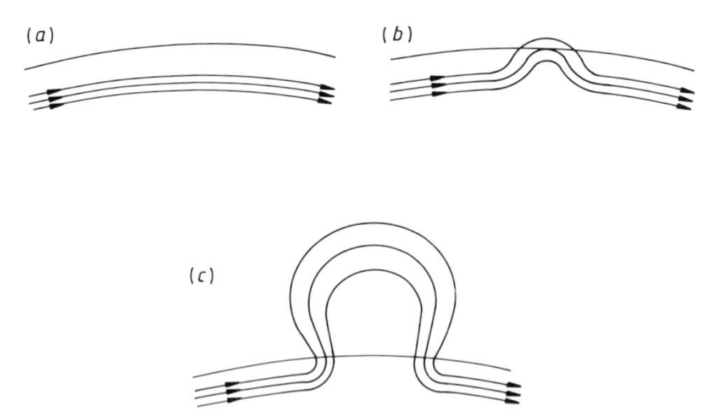

Figure 4.12 (*a*), (*b*) (*c*) The formation of sunspots and a loop prominence from a tube of force.

Over more complex sunspot groups the magnetic field can give rise to much more violent releases of energy. This can happen in the way shown in figure 4.13. When the field lines cross at a neutral point A they break and reconnect in such a way that they resemble the stretched elastic band of a catapault. These lines quickly snap back, accelerating particles both upwards towards the corona and downwards to the base of the chromosphere. The energy of the magnetic field is in this way converted to energy of motion of the particles. This energy can in turn be converted to radiation energy, giving rise to the complex emission phenomena seen during a flare event. One mechanism by which this can be done is the synchrotron process which was described in Chapter Two. The wavelength range in which most of the emission will occur depends on the strength of the magnetic field and the energy of the electrons. The radio bursts emitted by the Sun are largely generated by this process.

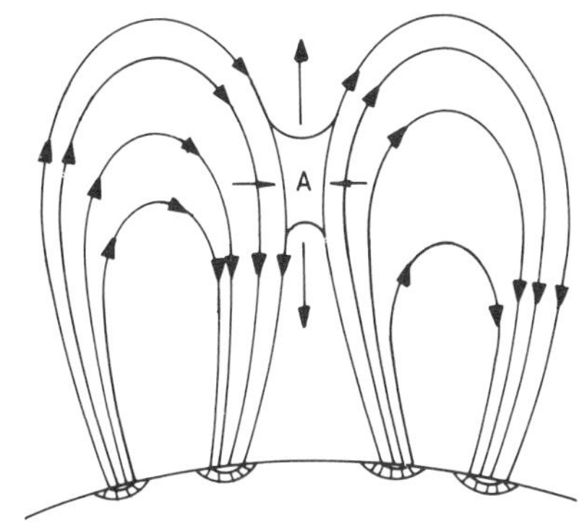

Figure 4.13 Magnetic loops over a complex sunspot group. Reconnection of magnetic lines of force is taking place at point A.

The Solar Wind and Interplanetary Magnetic Fields

The coronal holes mentioned earlier show up on x-ray photographs as dark regions, and are usually several times cooler than the active regions of the corona. Open magnetic lines of force diverge out of these holes, and eventually form part of the interplanetary field. Streams of particles follow the magnetic lines out of the holes and these particles form the solar wind. The evolution of these holes has been studied in some detail. A hole usually starts as a small hole near the equator, which then grows in size until it links up with the polar region of the same magnetic polarity. Thereafter the hole will begin to shrink in size—the total sequence of events will last about eight months.

The loops of the solar magnetic field which arch into the corona and are anchored in the active regions of low latitudes will be stretched out, near the ecliptic, into interplanetary space by the outflowing solar wind. While these loops are pulled out radially the Sun rotates about its axis, leading to the winding up of the stretched out field lines. As a result the interplanetary field near the ecliptic assumes a spiral configuration.

This is called the garden hose effect, because the lines of the magnetic field, like the water jet of a rotating garden hose, form a curved spiral, but the solar wind particles, like the droplets of water, always move in a radial direction (see figure 4.14). Although the particles in the solar wind move away from the Sun radially, an observer on Earth has the impression that they are coming from a direction approximately 5° to the west of the Sun. This is due to an aberration effect which is produced by the motion of Earth at right angles to the Sun–Earth direction. The effect is the same as the case of a moving observer who thinks that the rain is falling at an angle, while a stationary observer sees the rain falling vertically (figure 4.15).

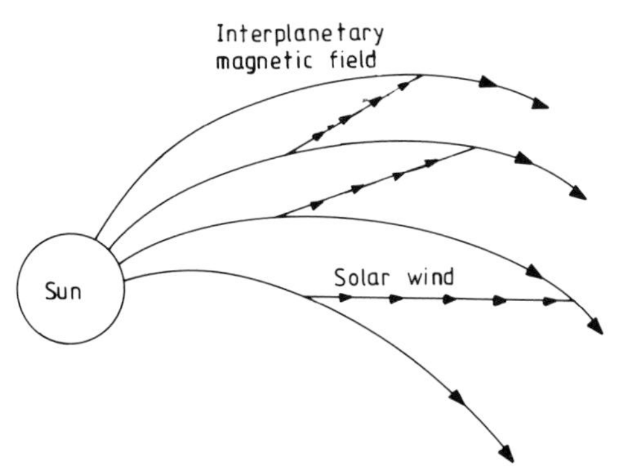

Figure 4.14 The solar wind and the interplanetary magnetic field.

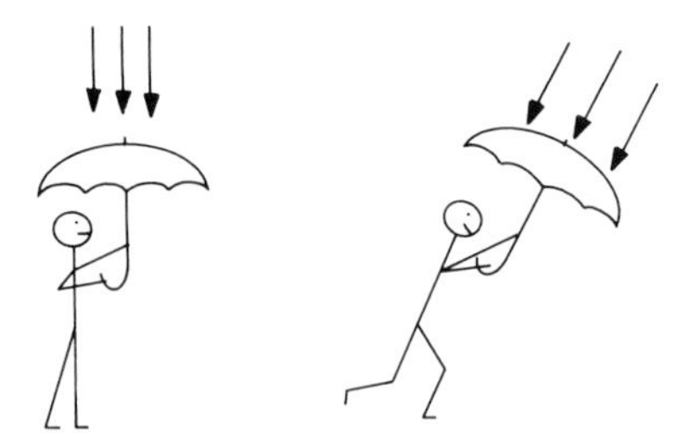

Figure 4.15 The aberration effect.

Besides having a spiral form the interplanetary field is divided into four sectors, with the field pointing in different directions in alternate sectors (figure 4.16).

Associated with the magnetic field there is a sheet of current. The warping of this sheet is due to the sectorial division of the solar magnetic field in the equatorial region of the Sun. A three-dimensional model of the sheet is shown in figure 4.17. The component of the interplanetary field which is perpendicular to the ecliptic, changes sign when the Earth passes through the current sheet as it orbits the Sun. This can lead to a reconnection between the lines of the interplanetary field and the geomagnetic field in the northern hemisphere, though not in the southern hemisphere where the interplanetary field has a downward component. When this happens charged particles can more easily enter the Earth's ionosphere, because charged particles preferentially travel along field lines. Geomagnetic activity thus changes significantly depending on which sector the Earth is in.

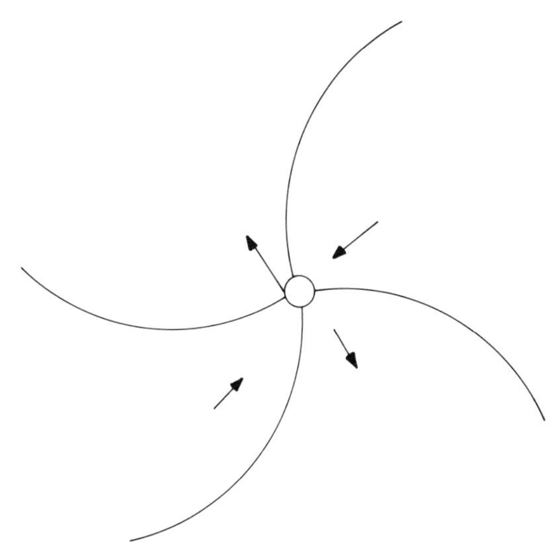

Figure 4.16 Sector boundaries of the interplanetary field.

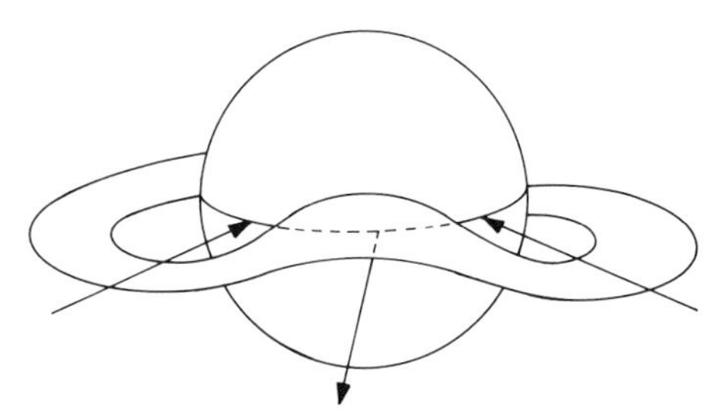

Figure 4.17 Three-dimensional model of the current sheet.

Since general solar activity increases considerably with increases in the number of sunspots, it is not surprising that the strength of the solar wind is also modulated by sunspot numbers. When sunspot activity is at its maximum there is an increase in the width of the auroral oval and a consequent increase in the frequency of aurorae in lower latitudes on Earth.

Activity on the Sun can also modulate the charged cosmic ray particles which come from beyond the solar system. One well known method by which this can occur is called the Forbush decrease. Gold and Parker both proposed theories to account for this sudden decrease in cosmic ray flux, which follows a large flare event on the Sun. According to Gold a plasma cloud ejected by a large flare event can pull out the solar magnetic field to form a large magnetic 'bottle'. Such a 'bottle' can eventually engulf the Earth and provide an additional shield from cosmic ray particles (figure 4.18). Parker suggests that a large flare event gives rise to a blast wave (similar to a sonic boom) in which the strength of the interplanetary field is increased. This blast wave propagates out into space and when it reaches Earth it provides additional protection against cosmic rays. The abrupt increase in the interplanetary magnetic field within such a blast wave has been observed by some spacecraft, thus providing support for Parker's theory (figure 4.19).

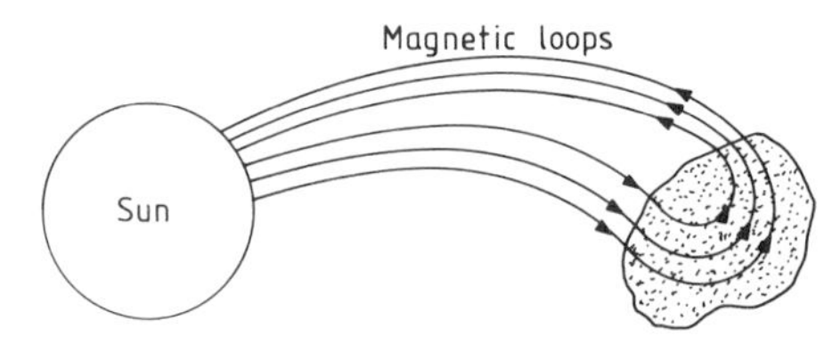

Figure 4.18 Gold's model to explain the Forbush decrease.

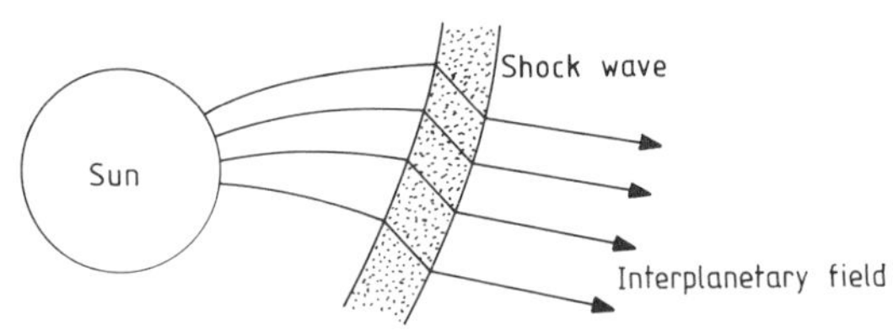

Figure 4.19 Parker's model for the Forbush decrease.

Possible Planetary Effects on Solar Activity

Since the mid 1940s there has been a growing realisation that the occurrence of particularly large solar flares is somehow associated with particular planetary configurations as seen from the Sun. Much of this started with the work of a radio engineer, Dr John Nelson, who at the time was employed by the RCA Company in America. He was given a project to find a way of forecasting the occurrence of severe increases in solar activity, since it was known that such increases could disrupt radio communications. Nelson discovered that when Venus, Earth, Mars, Jupiter and Saturn were almost in a straight line with the Sun, or when they made 90° angles with each other, as seen from the Sun, then conditions were likely to be particularly bad for radio reception. Conditions were good when the angles between planets, as seen from the Sun, were either 60° or 120°.

Dr Paul Jose, working for the US Air Force, was able to find a relationship between the maxima of solar activity and the movement of the Sun about the common centre of mass of the solar system. The work of Nelson and Jose was carried further by Drs Blizard and Sleeper, both of whom undertook projects on solar activity prediction on behalf of NASA. The methods of prediction were based on the alignment of planets with respect to the Sun, and although some astronomers would dismiss such an approach, their work did lead to some successful predictions of particularly violent solar events. Professor Wood, of the University of Colorado, was able to show that there was significant correlation between the peak periods of solar activity and the peaks of the combined tidal effects on the Sun due to the planets Venus, Earth and Jupiter. Wood ignored Mercury because its influence changes over a matter of days, and on the timescale of solar cycles its effect is more or less constant. The other planets were ignored because their calculated tidal effects were small in comparison to those included in the calculation.

There are two possible ways in which the planets could affect solar activity. First, the Sun's movement around the common centre of mass of the solar system could cause changes in the pattern of the convective motions which generate the large-scale solar magnetic field, and this in turn could lead to the large-scale reversals of the field. Tidal action could then foster the growth of instabilities in those regions where the magnetic field configuration is in unstable equilibrium. These suggestions in no way conflict with the earlier description of how solar activity arises from the behaviour of solar magnetic fields. The earlier description is really based on a theory that at present is still incomplete, and Professor Parker—an authority on cosmic magnetic fields—believes that a fully developed theory is still several years away.

Solar Activity, the Earth's Rotation, and Earthquakes

Dr R Challinor, of the University of Toronto, was one of the first scientists to suggest that there may be a link between events on the Sun, small changes in the rate of Earth's spin, and earthquakes. His suggestion was taken further by John Gribben who in 1971 made the tentative suggestion that 'One might even speculate that the San

Andreas Fault . . . might be triggered in this way in the late 1970s or early 1980s, shortly after the next period of maximum solar activity.' Gribben was later joined in his theorising by Stephen Plagemann and together they developed a theory on which they based a more precise prediction, that there would be a solar maximum in 1982, which would lead to a global series of earthquakes and that California in particular would be affected. The failure of their prediction should not be taken as a failure of the theory. As they are both now ready to admit, their enthusiasm for the general idea made them push some of the established correlations rather harder than was wise.

In the first place, although, as we have seen, there are links between positions and movements of the planets, as seen from the Sun, and solar activity, a foolproof theory of any sort, capable of predicting with great accuracy the next solar maximum, does not yet exist. Second, although there are indications that solar activity is related to the spin rate of Earth, the physical mechanism for this link is not understood. Third, there is evidence that changes in the spin rate of Earth do correlate with global increases in seismic activity, but it is impossible to tie this down to a given point on the Earth's surface.

Some of the most convincing evidence for links between the sunspot cycle and changes in the Earth's rotation rate comes from the work of Robert Currie, of the NASA–Goddard Spaceflight Centre in Greenbelt, Maryland. He used a very sophisticated computer technique to show clearly the influence of the solar cycle on small changes in the length of the day. The result can be partially understood if the solar cycle, by modulating the solar wind and the interplanetary magnetic field, causes changes in the circulation of the upper atmosphere, thus resulting in the change of spin rate of the Earth. Convincing evidence that the level of seismic activity at different times correlates well with changes in day length came from the work of Professor Don Anderson of the California Institute of Technology. Professor Anderson explained this result as follows. Seismic activity would give rise to increased volcanic activity. The dust thrown up by volcanoes alters the amount of solar radiation reaching the lower atmosphere, thereby causing changes in the circulation of the atmosphere, and these changes in circulation affect the spin rate of Earth.

The mechanism suggested by Gribben and Plagemann, in which solar activity affects spin rate and spin rate affects seismic activity, is more plausible than the one suggested by Anderson.

Solar Activity, Weather and Climate

Several researchers have claimed to have found correlations between solar activity and our climate and weather. One such claim seemed to show a link between the 22-year sunspot cycle and drought over the Western United States. Another claimed to have evidence for a correlation between atmospheric turbulence and the times at which sector boundaries of the interplanetary field crossed the magnetosphere. Most of these claims need more research before they can be either accepted or dismissed.

A more convincing theory about a longer-term climatic effect rests on the evidence for a link between the sunspot cycle and the rate of production of carbon-14 in the atmosphere. The carbon-14 in tree rings has been measured to show that solar activity was weaker in certain centuries than at other times. Using geophysical data on climatic conditions in the past it seems as if several periods of severe winters have corresponded to times when solar activity was much weaker.

Concluding Comments

An understanding of solar activity, solar magnetism, the solar wind and the interplanetary magnetic field is important if astronomers are to understand fully how our Earth and the other planets of the solar system are affected by the Sun. Because the Sun is so close to Earth its magnetic activity can be studied in more detail than that of any other star. These studies have shown that there are strong links between magnetic fields on the Sun and the radiation it emits at a large number of different wavelengths. These links are very important in investigating activity cycles of other sun-like stars. Knowledge of solar magnetic activity forms a model for investigating stellar magnetic fields.

Chapter Five

Magnetic Fields in the Solar System

This chapter looks at the magnetic fields of those planets that are known to have internally produced fields, at the magnetism of the Moon, and the solar wind-induced fields of some planets and around comets. The possibility that strong magnetic fields may have played a part in the early solar system is discussed and the evidence to support this briefly examined. Finally some general comments are made on the dynamo problem. However, since the internal structures of planets and their satellites are important to any discussion of the origin of their magnetism, the chapter starts by considering some general methods used to investigate planetary interiors and the results obtained. The spinning of the planets on their axes is a necessary condition for the generation of internally produced fields resulting from a dynamo mechanism, so a brief discussion of the rate of spin of each planet is also given.

Methods of Investigating the Internal Structure of Planets

Any discussion of the interior of the planets must start with a consideration of their mean densities, which in turn is based on a knowledge of their sizes and masses. Planetary diameters can be calculated from the angular sizes and known distances. The masses of planets with satellites can easily be calculated from the orbital periods of the satellites. For those planets which do not have satellites—i.e. Venus and Mercury—

their masses can be deduced from how far they affect, or perturb, the orbits of the other planets.

The mean densities alone cannot give any information on how the densities vary with distance from the centre. If the planets were exactly spherical in shape then the number of possible models that could give the observed mean densities would be very large. Fortunately the planets all have slight equatorial bulges, which results from the modification of their shapes by their rotations. This polar flattening can be used to obtain information on planetary interiors.

The variation in density with distance from the centre in the interior of a planet is related to quantities called the moments of inertia of the body. The moment of inertia is a measure of the planet's resistance to any changes in its spin about a given axis. Since the planets spin about their polar axes the polar moment of inertia of each planet is important. However, for the purposes of studying planetary interiors it is also important to know the moment of inertia about an axis at right angles to the polar axis.

In Chapter Three we saw how the Earth's axis precessed due to the combined gravitational pull of the Moon and Sun on the equatorial bulge. For the Earth it is possible to use the rate of precession, equatorial radius, Earth's mass and the rate at which it is spinning on its axis to calculate the ratio of equatorial to polar moments of inertia. The difference between these quantities can be calculated from the effect it has on the orbits of Earth satellites. The ratio and difference of the two moments of inertia can then be used to put constraints on models of the interior of Earth. Unfortunately it is only for Earth that a precession rate has been measured, so this particular method cannot be applied in this form to the other planets.

Although the Moon does not precess it does oscillate slightly about its own axis. It is normally stated that, on the whole, the Moon keeps the same face towards Earth because it rotates on its axis, with respect to the stars, once every 27.32 days, which is very nearly the same as the time it takes to go around Earth. This is only true to a first approximation. The Moon does oscillate about its axis, and this oscillation, called its physical libration, is due to the gravitational pull of Earth on the tidal bulge of the Moon. Since the Moon has an elliptical orbit about Earth, the tidal bulge in its crust does not always lie along the line from the Moon to Earth. The physical librations arise when the gravitational force of Earth tries to restore the bulge to this line, thus setting up an oscillation about the Moon's axis. These librations can be used, in the same way as the precession of Earth's axis is used, to give information on the interior of the Moon. The libration measurements must, however, be combined with information on the gravitational field of the Moon obtained from the orbits of spacecraft.

In order to use the methods described above for other planetary interiors it is necessary to employ more theoretical considerations. For very large masses like the planets, gravity is the dominant internal force, and over long periods it can mould the solid parts of a planet's interior just as if the planet were made of liquids. This means that conditions of hydrostatic equilibrium prevail in planetary interiors. This equilibrium is, however, modified by the rotation of a planet, and the modification can be used to calculate the polar moment of inertia from the mass, equatorial radius, rotational periods, and measured polar flattening or knowledge of the external gravitational field obtained from natural satellites or spacecraft. It is now possible to describe the results obtained from these methods.

Interiors of Jupiter and Saturn

The main features of Jupiter's interior are shown in figure 5.1. The core of Jupiter is probably rocky, consisting of iron silicate. This is probably surrounded by a thick shell of liquid metallic hydrogen plus traces of helium. The outer thinner shell is composed of liquid molecular hydrogen containing some helium.

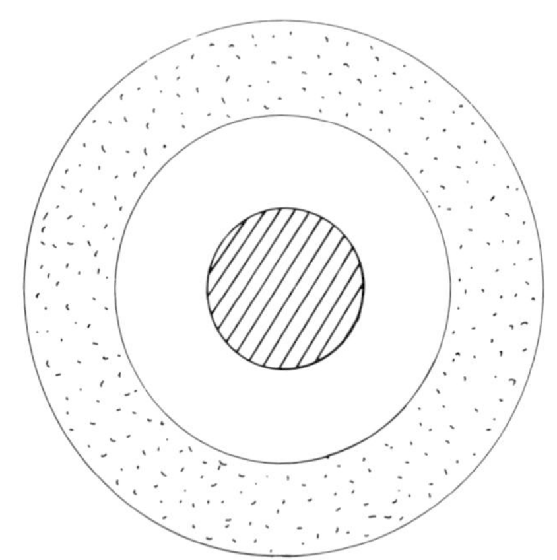

Figure 5.1 Internal structure of Jupiter. A small rocky core surrounded by a shell of liquid molecular hydrogen and an outer shell of molecular hydrogen.

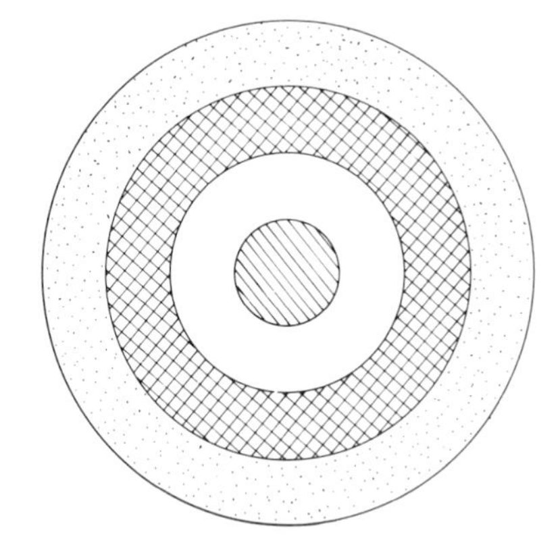

Figure 5.2 Internal structure of Saturn. A rocky core surrounded by a shell of ice. This is enclosed by a shell of metallic hydrogen with an outer shell of molecular hydrogen.

Saturn consists of a rocky core surrounded by a thin shell of metallic hydrogen and helium (figure 5.2). The outer shell, of molecular hydrogen and helium, occupies a larger proportion of the volume of Saturn than does the outer shell in the case of Jupiter. The models for the two planets as described are not considered final, and will probably undergo revision as we learn more about these planets.

Internal Structure of Terrestrial Planets and the Moon

Mercury consists of a silicate mantle surrounding an iron core. As will be seen later, Mercury does have a magnetic field which would seem to suggest that at least part of the core may still be molten. If this is not the case then the presence of a magnetic field of internal origin will present a major problem (figure 5.3).

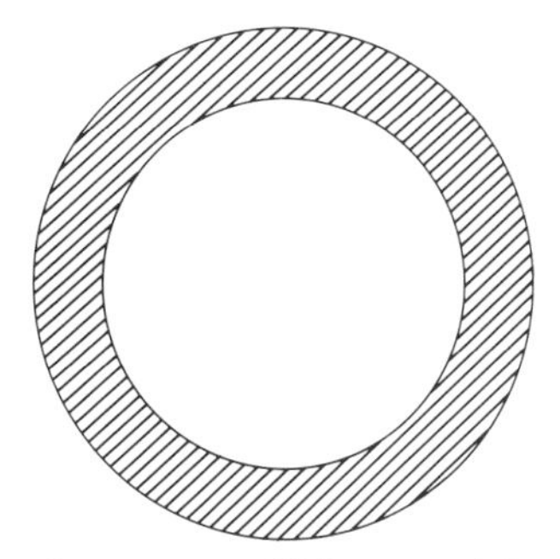

Figure 5.3 Internal structure of Mercury. A large core surrounded by a relatively small mantle and a thin crust.

The size and mass of Venus is very similar to that of Earth, giving an average density only slightly less than that of our own planet. It is believed that the internal structure of Venus is probably similar to that of Earth with an iron core and a mantle of silicate and oxide.

For the Moon the methods described above have been supplemented by seismic studies. Astronauts placed seismic instruments on the lunar surface which recorded lunar quakes, including some induced by crashing redundant parts of spacecraft onto the Moon's surface. These studies showed that the Moon has a dense core of radius 500 km and a mantle 1200 km thick. The mantle is covered by a thin crust.

Rotation of the Planets

When discussing the rotation of the planets about their own axes it is normal to quote the sidereal rotation period. This is the rotation measured with respect to the stars rather than with respect to the Sun.

Measurements of the rotation rate of Mars present no special problems since Mars has a clearly distinguishable surface. The rotation rates of the four other planets visible to the naked eye ('naked-eye planets') do, however, present special problems.

Jupiter and Saturn present problems because they probably do not have a solid surface and it is only the upper features of their cloud systems that are visible at optical wavelengths. There is also the added complication that features at different latitudes rotate with different speeds. For Jupiter the first step towards solving these problems was to define two separate systems of rotation. System I rotates faster than the rest of the planet. It extends to 10° north and south of the equator and has a rotation period of 9 h 50 min 30.003 s. The rest of the planet, from 10° north up to the north pole and from 10° south down to the south pole, constitutes System II, which has a rotation period of 9 h 55 min 40.062 s. These periods of rotation are used to define two systems of longitude. In System I the longitude of the central meridian increases by 36°.579 per hour and in System II the longitude increases by 36°.262 per hour. Radioastronomical observations revealed that prominant radio features of Jupiter oscillated with yet another period. This period has been measured as 9 h 55 min 29.75 s, and it is used to define System III which, though detectable at radio wavelengths, is not visible.

In addition to the problems already mentioned for Jupiter, Saturn presented a further difficulty. This is due to the fact that clearly distinguishable features at different latitudes are not easily visible with Earth-bound optical telescopes. Mercury and Venus also had their own peculiar problems, Mercury because it could only be seen at specific points on its orbit around the Sun, and Venus because of its thick cloud cover. All these separate problems were overcome using radar astronomical techniques.

In radar astronomy radio signals are transmitted to a planet by a radio telescope and the same telescope receives the returning signal which has been reflected back from the planet. The transmitted signal is of a definite known frequency, but the return signal from different parts of the planet will be Doppler-shifted by different amounts with respect

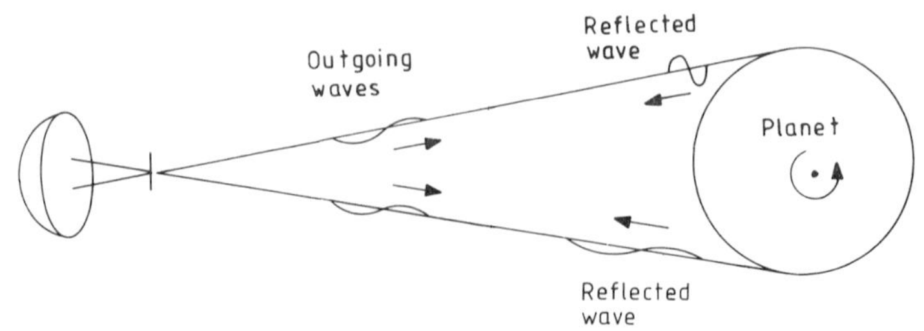

Figure 5.4 Using radar to determine planetary rotation rates.

to the transmitted signal. The difference between the transmitted signal and that returned from the centre of the planetary sphere can be used, according to the Doppler effect, to find the speed with which the planet and the Earth are approaching or receding from each other. The difference in frequency, due to Doppler shifting, between the two limbs of the planet can be used to calculate its rotation rate (figure 5.4).

When the radar techniques were applied to Saturn they showed the following relationship between latitude and rotation period:

Latitude	Rotation period
0° (Equator)	10 h 2 min
27°	10 h 38 min
42°	10 h 50 min
57°	11 h 8 min

The application of these techniques to Venus gave the surprising result that Venus has a rotation period of 243 solar days, but in the opposite direction to that of other naked-eye planets. For Mercury the method gave an axial period of 58.64 days.

Having discussed the internal structure and rotation rates of the five naked-eye planets and the Moon we can now go on to discuss their magnetic fields.

Magnetospheres of Jupiter and Saturn

Ever since the early days of radioastronomy in the 1950s it has been known that Jupiter was a strong emitter of radio waves. These observations led to the discovery that Jupiter has a strong magnetic field. The radio maps made at 10 cm wavelengths showed that most of the radiation did not come from the planet itself, but from two lobes on either side of the planet (see figure 5.5). More detailed observations showed

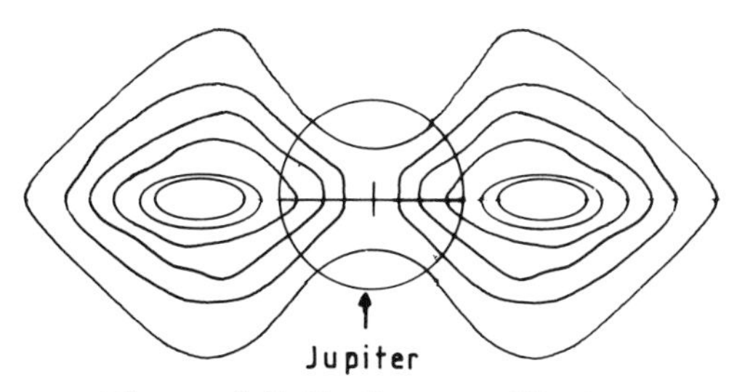

Figure 5.5 Radio map of Jupiter.

that the intensity of the radio radiation, its linear and circular polarisation all varied with a period equal to that of System III. This led to the theory which attributes the radiation to the synchrotron effect operating on high energy electrons spiralling in the dipolar magnetic field of Jupiter. This theory made it possible to deduce the angle between the magnetic axis and the rotation axis. The relationships between the orientation of the dipole, the intensity and polarisation (both linear and circular) is shown in figure 5.6. The angle between the dipolar axis and the rotation axis is 10.77°, and the centre of the dipole is 0.101 times the planet's radius from the centre of the planet. A terrestrial compass would point to the south pole of Jupiter and the maximum surface strength of the field is about 20 times greater than that of Earth.

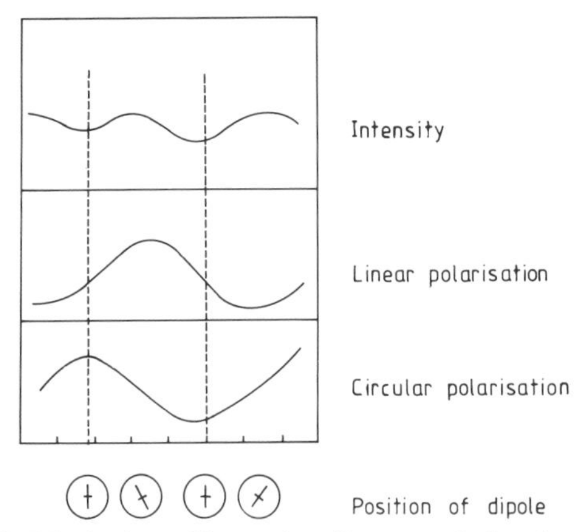

Figure 5.6 Variation of intensity, linear and circular polarisation with orientation of the Jovian dipole.

The radioastronomical observations of Jupiter's magnetosphere have been confirmed and considerably extended by data gathered by magnetometers and other equipment on board the Pioneer 10 (1973), Pioneer 11 (1974), Voyager 1 (1979) and Voyager 2 (1979) spacecraft. The observations have shown that the Jovian magnetosphere is the largest object in the solar system.

The Jovian magnetosphere can be defined as the region in which the magnetic field of the planet dominates over the interplanetary one, and it consists of three distinct parts (figure 5.7). The results obtained from Voyager flybys show that the inner magnetosphere is dominated by the dipolar field out to a distance of about six Jovian radii, which corresponds with the orbit of Io. In this inner region the magnetosphere behaves in much the same way as the magnetosphere of Earth. Photographs of the night side of Jupiter (see figure 5.8) show auroral discharges in its upper atmosphere.

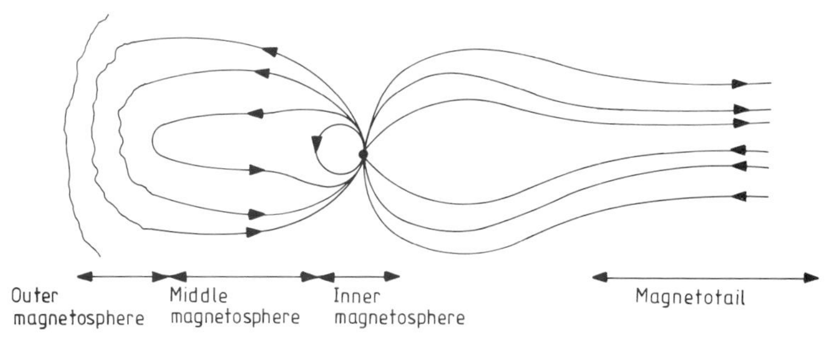

Figure 5.7 The inner, middle and outer magnetosphere of Jupiter.

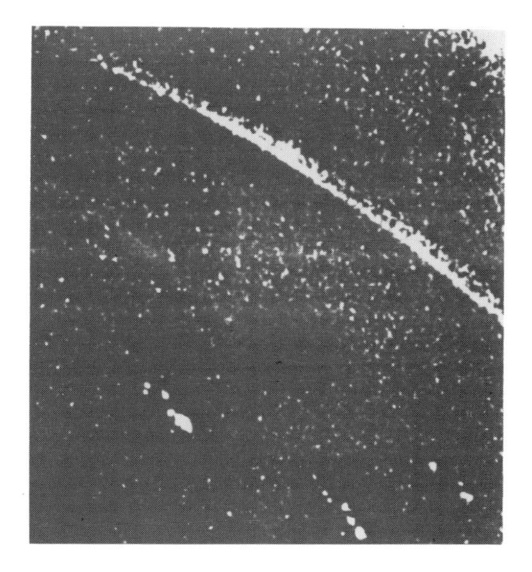

Figure 5.8 The white streak represents an auroral discharge, and the white specks near the bottom of the photograph represent lightning flashes. (Courtesy JPL/NASA.)

Other Voyager observations showed that Whistlers, similar to those on Earth, also exist in the Jovian magnetosphere. The middle magnetosphere extends from six Jovian radii to about 30–50 Jovian radii from the planet. This is a thin disc-like region in which the magnetic field is nearly constant. It contains a low energy plasma, in which there is a current sheet. The tilt of Jupiter's magnetic axis warps this sheet of current so that it is above the equator on one side and below it on the other side (figure 5.9). The current sheet and the magnetic field rotate with the planet. The field of the outer magnetosphere extends from the boundary of the middle magnetosphere to the magnetopause, and includes the extensive magnetotail of the planet. Normally this field has a strong southward component but it also exhibits large spatial and temporal changes in direction and magnitude in response to changes in the pressure of the solar wind.

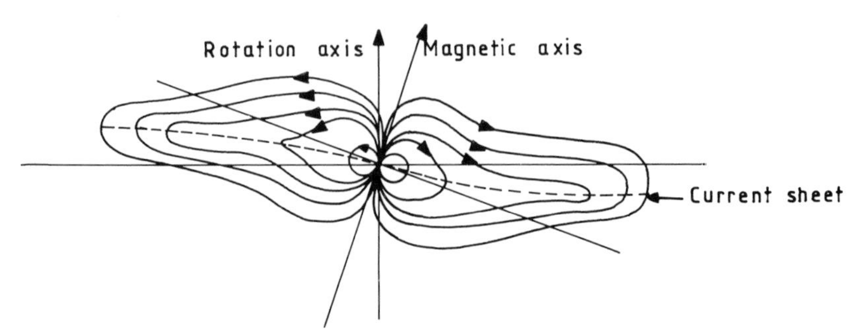

Figure 5.9 The middle magnetospheric disc of Jupiter.

There is a close interaction between Io, the innermost of Jupiter's Galilean moons, and the inner magnetosphere of Jupiter. There is substantial evidence for two effects of this interaction, and the possibility of a third form of interaction has been suggested.

The effect of Io on radio bursts from the planet at wavelengths between 670.0 and 7.5 m has been known of for some time. An important feature of these bursts is that they occur mostly when the longitude of the central meridian of System III has certain values. This seemed to imply that the bursts are associated with certain preferred zones of longitude. Three such zones were identified and they were labelled A, B and C. Bursts from A and B are most likely to occur when Io is at certain positions of its orbit (see figure 5.10). Although Io plays the most

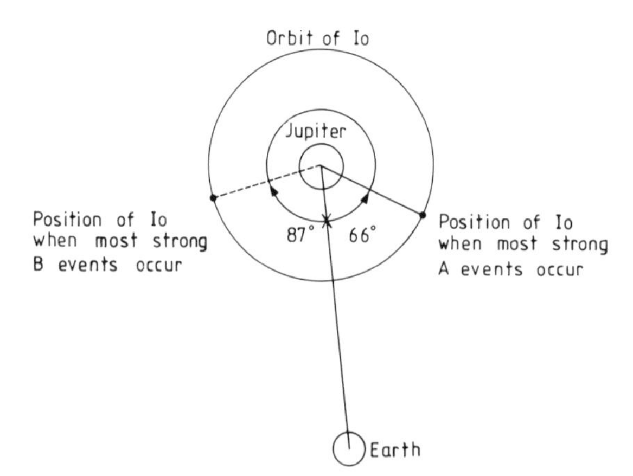

Figure 5.10 The relative positions of Earth, Io and Jupiter during certain radio bursts.

important part in these bursts, the satellite Europa is also involved to some extent. The bursts are the result of an interaction between the inner magnetosphere of Jupiter and the ionosphere of Io. They are believed to be due to very strong electric current flows caused by Io 'closing the switch' of an electric circuit. As Io moves through the magnetosphere its ionosphere becomes positively charged on one side and negatively charged on the other, thus giving rise to an enormous electrical potential across the satellite. When Io reaches a certain point in its orbit, current flows along magnetic field lines down to Jupiter's conducting ionosphere, and then back to Io, thus completing the electrical circuit (figure 5.11).

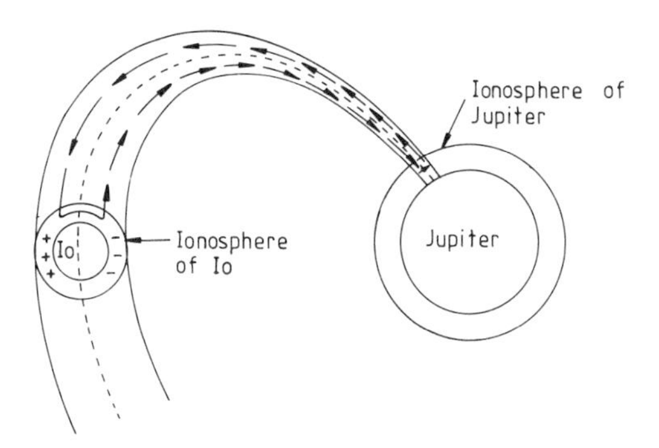

Figure 5.11 Io's effect on radio noise from Jupiter.

It is not yet clear how this current is converted into radio waves, but a possible theory is being developed. This theory involves magnetic waves travelling along the field lines involved with the electrical circuit. In Chapter Two it was stated that field lines behave, in certain respects, like stretched elastic bands, so it is possible to set a field line vibrating rather like the string of a musical instrument. The waves set up in this way are called Alfven waves. The theory which tries to explain the radio burst assumes that the completing of the electrical circuit between Jupiter and Io could lead to Alfven waves being set up in some of the field lines, and that these waves could accelerate the electrons to high energies at which they could generate radio waves. Although this is a promising theory, the details are not yet all clear and more research is necessary before a full explanation of the Io effect can be given.

Interaction between Io and the Jovian magnetosphere may explain another problem, i.e. the origin of the energy source (or sources) necessary to provide large internal heating of the satellite. Photographs of Io taken by cameras on board the Voyager spacecraft show that volcanic calderas, with and without flow structures, cover large areas of its surface (see figure 5.12). Some of the photographs actually showed volcanoes erupting (figure 5.13) thus confirming that Io is the most

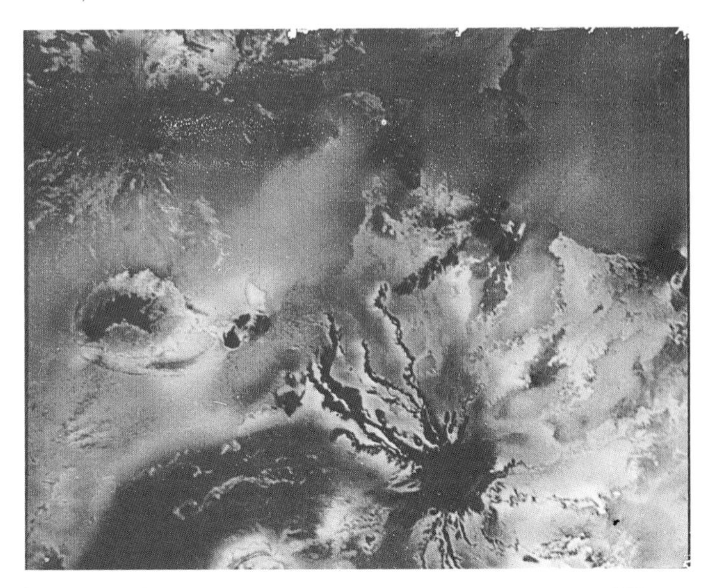

Figure 5.12 Lava flows on Io from 128 500 km. (Courtesy NASA/NSSDC.)

Figure 5.13 Volcanic explosion on Io from 490 000 km. (Courtesy NASA/NSSDC.)

volcanic body in the solar system. From the heights of the volcanic plumes it is possible to calculate the ejection velocities of the volcanic materials. These calculations show that typical ejection velocities are about ten times greater than those for terrestrial volcanoes. This has led to the theory that the volcanoes are powered by 'geysers' in which sulphur dioxide is the principal propellant.

All the observed volcanic activity raises important issues concerning the source of energy necessary to keep the interior of Io molten. Simple models of the thermal history for satellites and small planets are not consistent with the temperatures necessary to keep Io's ingredients in a fluid state. Some heat would have arisen from the accretion of the materials which formed the satellite, and some heat will have resulted from the radioactive decay of short-lived radioactive substances. However, this heat would have been lost rather quickly, and most other sources of heat, such as long-lived radioactive materials, are not as effective over long timescales.

Another possible source of heating is the tidal deformation of Io's crust by the satellite Europa. This is the result of a phenomenon known as gravitational resonance, which exists between the two satellites. Io has an orbital period of 1.769 days compared with 3.551 days for Europa. This means that Io feels a gravitational tug from Europa every two of its own revolutions. As a result Io is forced into an eccentric orbit. Just as our Moon, to a first approximation, keeps the same face towards the Earth, so—also to a first approximation—Io keeps the same face towards Jupiter. However, the Europa-induced eccentricity of its orbit makes Io travel at different velocities along its orbit. As a result the Jupiter-induced tidal bulge of Io rocks back and forth slightly as seen from the planet. However, the bulge wants to move with respect to the satellite, so that it lies along the line joining its centre to the centre of Jupiter. This movement causes deformation of Io's crust and leads to internal heating of the satellite.

A further possible mechanism for heating Io's interior could arise from its motion through the inner magnetosphere of Jupiter. It has already been seen that this movement causes a large electrical potential difference across the diameter of the planet. Internal currents resulting from this potential could give rise to the heating of Io's interior. The volcanoes on Io are probably one of the sources of the plasma torus surrounding Jupiter at the orbit of Io.

In 1974 Robert Brown discovered sodium emission from the neighbourhood of Io. Later observations showed that the emission was coming from a 'cloud' of neutral sodium surrounding Io. This cloud extended a considerable distance along Io's orbit in the shape of a banana. Potassium and ionised sulphur were also detected in this cloud shortly afterwards. All these discoveries were made from terrestrial observatories. However, it was as a result of measurements made by Voyager 1 that the full extent of Io-associated atomic particles and ions in Jupiter's magnetosphere became clear. These particles appear to originate in the plasma torus at Io's orbit, and it seems very likely that the ultimate origin of the particles is the satellite itself. However, it is not yet clear how these particles can escape from the gravitational pull of Io. Not even the particles from the volcanoes have enough energy to overcome the gravitational attraction.

During their encounters with Saturn the Voyager spacecraft detected and measured the properties of its magnetic field and its magnetosphere. Saturn also has a dipolar magnetic field but unlike Earth, Jupiter or Mercury, the axis of the dipole is very close to the rotation axis, being inclined to it by about 1°. The dipole is also situated closer to the centre than is the case for the other planets. This presents problems for a dynamo origin of the field, which predicts a dipole displaced from the centre, and inclined to the rotation axis at an angle of greater than 1°.One possible explanation for these discrepancies between theory and observation is that the Saturnian field is about to reverse its polarity, just as Earth's field has done in the past.

The general structure of the magnetosphere of Saturn is shown in figure 5.14. It contains neutral gas and ionised plasma. The two satellites Titan and Rhea orbit the planet within the magnetosphere. Close to the planet there is one torus consisting of molecular hydrogen gas and further out there is another, much larger, torus consisting of atomic hydrogen. the plasma consists of an inner sheet, between these two hydrogen toruses, and an outer sheet, which is compressed on the sunward side by the solar wind, but extends for a considerable distance on the other side. Unlike the magnetosphere of Jupiter, Saturn's does not give rise to radio radiation which is detectable from Earth.

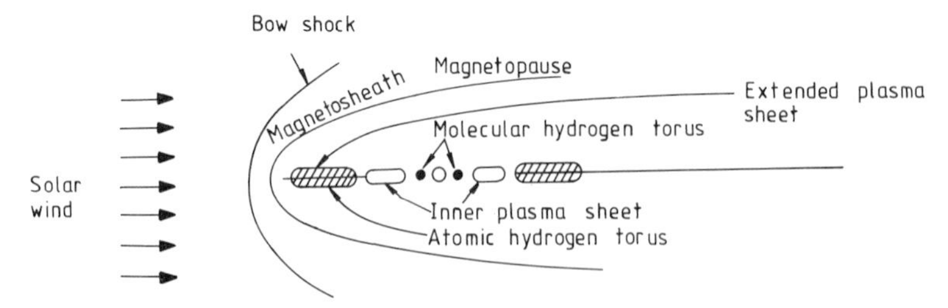

Figure 5.14 The magnetosphere of Saturn. (Reproduced with permission from *The Solar System* by B W Jones (1984) (Oxford: Pergamon).)

The magnetic fields of Jupiter and Saturn are believed to be due to dynamo mechanisms. Such dynamos could well exist in those regions of the planetary interiors of these bodies where there is metallic hydrogen. Although the near alignment of the rotation and magnetic axes of Saturn does present some problems for the dynamo theory there are ways of overcoming these problems. Some general comments on dynamo theories in our solar system will be made at the end of this chapter.

The Magnetic Field of Uranus

Towards the end of January 1986 the Voyager 2 space-probe encountered the planet Uranus and the results of the encounter revealed several surprises, including one concerning the nature of its magnetic field. However, before describing this particular aspect of the results, it is necessary to discuss a few relevant facts about the planet.

First we consider a recently developed model of the planet's interior. According to this model the central pressure of the planet is about twenty million times the pressure of our own atmosphere and the central temperature is about 7000 kelvin. The hot 'rocky' core is composed largely of iron and silicates, in either solid or liquid form. This is surrounded by a mantle of frozen water, methane and ammonia. The outer shell is a gaseous mixture of hydrogen and helium, the outermost layer of which forms the atmosphere observed from Earth. This shell might well be solid at its base. Uranus has a rotation period of about 16 hours 48 minutes and its axis of rotation lies almost in the plane of its orbit.

The detection of Uranus' magnetic field came later than expected. Radio emissions connected with the magnetic field were only picked up by Voyager 2 about five days before its closest approach to the planet. Definite evidence for the magnetic field came three days before closest approach, when the probe detected the 'bow shock'. The field is about 15% stronger than that of Earth. The major surprise was the large angle between the rotation axis and the magnetic axis. This turned out to be 55°, much more than any other planet. Two suggestions are being discussed which could account for this. One was that the field was due to fossilised permanent magnetism which for some reason had become frozen into the internal structure of the planet at this particular angle. Such a possibility is not unknown, as will be seen in a later chapter, in magnetic stars and pulsars, but it has not been observed in any other planetary context. The other possibility is that the atmosphere has a different axis of spin from the core. This could be explained if the planet had, sometime in the past, been struck by another body, of about the same size as Earth, which altered the spin axis of its atmosphere. This possibility had already been discussed by astronomers in connection with the fact that Uranus' spin axis is almost in the plane of its orbit.

Planetary Rings

Prior to 1974 Saturn was the only planet known to have a ring system. Now there is definite evidence that Jupiter has a ring, Uranus has a ring system, and recent observations suggest that Neptune may also have a system of rings. The suggestion that Jupiter may have a ring first came from Acuna and Ness. They noticed that as Pioneer 11 moved through Jupiter's magnetosphere in 1974, getting to within 1.6 Jovian radii of the planet's centre, counts of high energy particles dropped whenever a known satellite was passed. Such a drop in counts was noticed at a distance of 1.7 and 1.8 Jovian radii. Acuna and Ness suggested that this drop could either be due to an undiscovered satellite or to a ring. The presence of a ring showed up clearly in photographs taken by Voyager 2 (figure 5.15).

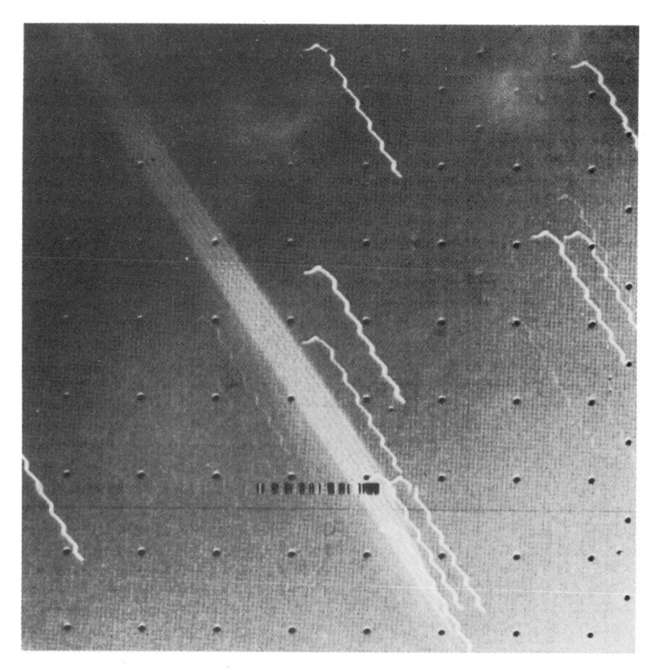

Figure 5.15 Evidence for the ring of Jupiter plus jagged star trails. (Courtesy JPL/NASA.)

Uranus' ring system was detected in 1977 as the planet occulted a star. It was noticed that nine minor occultations took place both before and after the main occultation of the star by the planet. These minor interruptions of the light from the star were due to the rings of Uranus passing in front of the star. A similar set of observations on Neptune has led to the suggestion that this planet also has a ring system.

Evidence from Voyager observations of the rings of Saturn suggest that its rings do interact with its magnetosphere. One of the surprises of Voyager 1's encounter with Saturn was the detection of a number of spokes or dark fingers in one of the rings of Saturn (figure 5.16). Ordinary orbital mechanics require that particles orbiting a body at different distances should travel at different speeds. Since the rings of all the planets consist of particles orbiting at different distances, any bunching of these particles to form spokes would seem to be destroyed

Figure 5.16 Part of Saturn's ring system. (Courtesy NASA/ NSSDC.)

by these variations in speed. Yet the observations showed that these spokes seem to be rotating with the planet. Gravity and the laws of motion on their own cannot explain this behaviour. However, an explanation is possible in terms of the interactions between the ring particles, the magnetospheric ions and electrons, and Saturn's magnetic field. The particles could well pick up electrical charges from the ions and electrons. The smaller particles would then have appreciable electromagnetic forces acting on them in addition to the force of gravity, as they move at right angles to the field lines. Various theories based on these considerations are being investigated at the moment.

Magnetic Measurements of Terrestrial Planets

Because of its comparatively small size and slow rotation rate, it was predicted that Mercury would have no intrinsic field except possibly for a small field due to permanent magnetism caused by the solar wind and the interplanetary field. Mariner spacecraft in 1974 and 1975 measured a field with a strength of about 1% of Earth's field, and showed it to be dipolar in nature. Two models of the strength of the dipole have been proposed which lead to two different estimates for the angle between magnetic and rotation axes. One gives a value of 2.3°, while the other gives a value of 14.5°. It has been argued by one scientist that Mercury's field arises from a thin shell dynamo maintained by chemical convection and latent heat released by the growth of the core. Another scientist has argued that it is due to remnant magnetisation.

Mercury also has a magnetosphere, although it is much smaller than that of Earth because its dipole field is weaker and, being closer to the Sun, the solar wind is stronger (see figure 5.17).

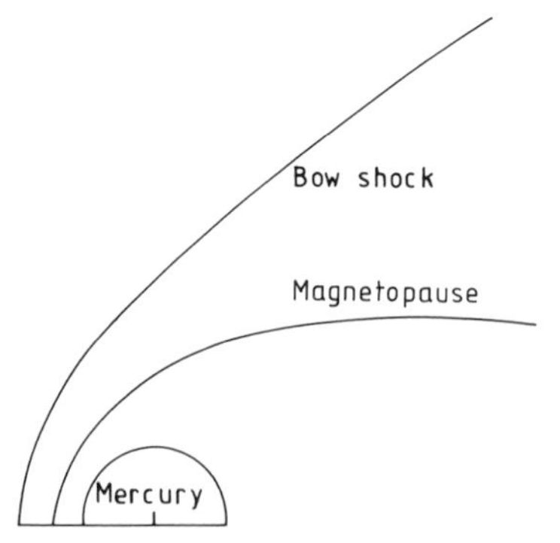

Figure 5.17 Bow shock of magnetopause of Mercury. (Figures 5.17 and 5.18 from the article by J A von Allen in *The New Solar System* (2nd edn) ed J K Beatty, B O'Leary and A Chaikin (1982) (Cambridge: Cambridge University Press).)

The magnetic environment of Venus was studied by a series of Mariner and Veneras space probes and also by the Pioneer Venus Orbiter. All these investigations showed that strength of the dipolar field of Venus is less than one ten-thousandth of Earth's field. This result is somewhat surprising since Venus probably has a liquid core and is comparable in size with Earth. The answer to the very low strength of the dipolar field may lie in Venus' very slow rotation rate. Another possibility is that no dynamo operates in Venus because it has a completely fluid core which is divided into stable strata, and as a result there is no adequate source of energy to power the dynamo. However, eddy currents induced in the conducting ionosphere of Venus prevent the solar wind from reaching its surface and as a result the planet has a well developed bow shock (figure 5.18).

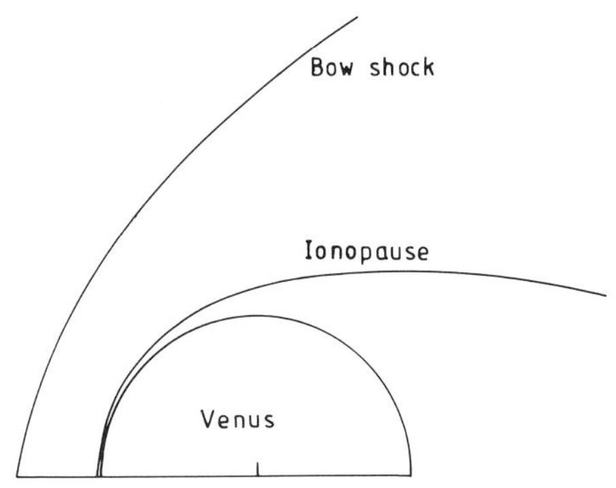

Figure 5.18 Bow shock and Ionopause of Venus.

The case of the magnetic field of Mars is still not yet clear. Russian astronomers have interpreted their results as showing that Mars has a dipolar field (figure 5.19) but the Americans have interpreted their own data as showing a field caused by the solar wind and the interplanetary field. More spacecraft measurements are necessary before the matter can be clarified.

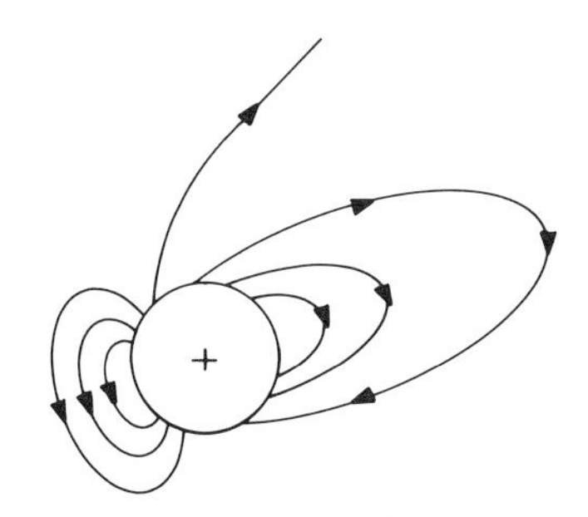

Figure 5.19 A Russian view of the magnetic field of Mars.

Lunar Magnetism

Three different methods have been used to study the magnetic field of the Moon. The first method uses laboratory-type magnetic detectors on board lunar probes or spacecraft. The second involves studying, by means of instruments on board spacecraft, the way the Moon's surface reflects charged particles contained in the solar wind. The third involves the study of lunar samples brought back to Earth by astronauts.

Some of the electrons contained in the solar wind can be reflected back from the lunar surface under certain conditions and these reflected electrons can be measured by particle detectors on spacecraft orbiting the Moon. This method is called electron reflectance magnetometry. When the Moon is in the magnetic tail of Earth then the lines of force of the tail will tend to bunch around those areas of the Moon where the magnetic field is particularly strong. At certain times when this situation arises there can be a direct connection between the interplanetary lines of force and the lines of the geomagnetic tail (see figure 5.20). Solar wind electrons can then stream along such lines to reach the dark side of the Moon, and some of these will be reflected from areas of the Moon with enhanced magnetic fields. By comparing the number of particles incident on the surface with the number reflected from various areas it is possible to map the strength of the magnetic field of the Moon. Unfortunately this method does not give information on the direction of the field, and so to obtain a more complete picture these results have to be combined with results obtained using the other two methods.

These studies have shown that the magnetic field of the Moon is due to widespread permanent magnetism. It has also been shown that there are magnetised lava flows on the Moon. However, at the moment the Moon does not appear to have an overall magnetic field of internal origin.

We have already seen that when substances are heated above their Curie temperatures they lose their magnetism, so any magnetism in lunar rock would have been destroyed when it melted to form lava. This implies that a field must have been present to magnetise the lava flows as they cooled. This raises the very challenging problem of the origin of the Moon's magnetic field at the time of the lava flows. Seismic evidence rules out a liquid core being present at the moment, but the possibility does exist that in the past the Moon may have had a liquid convecting core and, by analogy with Earth, this core could have generated a magnetic field and remnants of this field could have been present when the lava flows occurred. However, there is no other evidence to support the idea of a convecting core in the past, and there are also theoretical objections to the idea, so the problem remains to be solved. Nevertheless several possible explanations have been proposed in recent years.

One suggestion involves the production of a magnetic field by thermoelectric currents that could pass from a basin of magma (mineral matter associated with volcanoes) with a crust to a basin without a crust. Since the two will have different temperatures the current is completed above the ground by passing through any plasma that may be present. With such a mechanism it may be possible to produce some small-scale magnetic features but the problem still remains for the more extensive magnetic regions.

When a large meteorite crashes into the lunar surface it produces a shock wave that can form plasma. Strong magnetic fields associated with currents in the plasma could produce the observed permanent

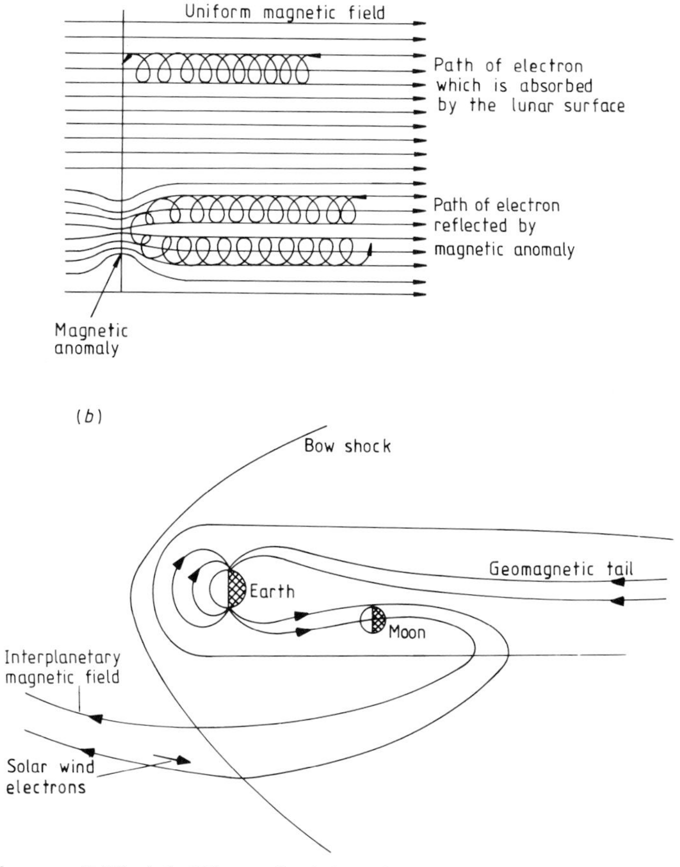

Figure 5.20 (*a*) The principle of lunar electron reflectance magnetometry. (*b*) Solar wind electrons reaching the Moon via a reconnection between lines of force of the interplanetary field and geomagnetic tail.

magnetisation. This process can produce magnetic fields of its own, but it can also magnify any existing fields. Once again it is difficult to see how the process can produce the observed coherence over large-scale magnetic features.

A third proposal involves a strong external field from the Sun. Many researchers have proposed that the Sun went through a T-Tauri phase of evolution. T-Tauri stars derive their name from the best-known member of the group, T-Tauri in the constellation of Taurus (the bull). They are pre-main sequence objects, very young on solar system time-scales, unusually bright, have very strong solar winds and associated magnetic fields. If the Sun did go through such a phase it would have had a much stronger field than it has now, and this field could have magnetised parts of the Moon. However, it is difficult to see how magnetisation produced so long ago could have survived to the present day. Part of this problem can be overcome by assuming that the initial field could have been amplified by plasmas produced in shock waves resulting from meteoritic impact.

It may be that several of the proposed mechanisms may have to be used in conjunction with each other to produce a result which matches the observed properties of magnetic fields on the Moon, or the solution may well lie in a different direction altogether.

Magnetic Fields in the Early Solar System

The proposed T-Tauri phase for the Sun has also been used to explain a problem associated with the origin of the solar system. This particular problem is concerned with the amount of angular momentum in the rotation of the Sun as compared with the amount in the orbital momentum of the outer planets. If the Sun and the planets were all formed out of the same interstellar cloud, then according to calculations the Sun should be spinning on its own axis much faster than observations indicate, and the motions of the outer planets around the Sun should be much slower. However, if the Sun had a much more powerful field during the T-Tauri phase, then this field could have transferred angular momentum from the Sun to the outer planets. This is because the magnetic field would have been rotating with the Sun but the moving magnetic field would have tended to move the electrically conducting planets in front of it. This would have led to a slowing down of the Sun but a speeding up of the planets (see figure 5.21).

If the Sun did go through a T-Tauri stage, then the enhanced magnetic field in interplanetary space would have been fossilised in meteorites formed at that time. Several meteorites have been examined for such magnetism and some of them yield magnetic measurements which seem to be consistent with such a proposal. However, there seems to be no consensus of opinion concerning the interpretation of these results, and more data are necessary if this question is to be settled unambiguously.

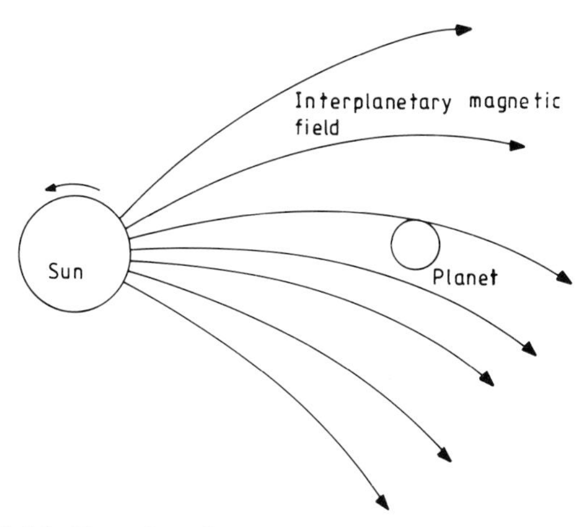

Figure 5.21 Transfer of angular momentum from the Sun to a planet via the interplanetary magnetic field of the early solar system.

Comets and the Interplanetary Field

The nucleus of a comet consists of a lump of compressed water-ice, impregnated with impurities, and covered with a layer of material resembling soot. As a comet approaches the Sun some of the water-ice is boiled off to form an atmosphere, known as the coma. When the comet is close enough to the Sun the solar wind blows some of the gas and dust of the coma into two tails—one of gas and one of dust. The tails have slightly different shapes since the dust is not as easily affected by the solar wind as the gas. Since the nucleus is solid, no convective motions can take place for the comet to generate its own magnetic field. Although comets do not have intrinsic magnetic fields, since a dynamo cannot operate in the nucleus of a comet, they do interact with the solar wind and the interplanetary field in a way that is not dissimilar to the behaviour of these entities in the neighbourhood of Venus. A schematic model for the interaction is shown in figure 5.22. In front of the outer shock wave the solar wind has a speed of about 400 km s^{-1}, but between the inner and outer shock the speed is reduced to 50 km s^{-1}. In this region the interplanetary magnetic field is very irregular. There is a tangential discontinuity separating the flow of the solar wind from the outflowing of plasma from the comet. In this region there is a bunching up of the interplanetary field, and here the strength of the field is about 10 times its value just in front of the outer shock.

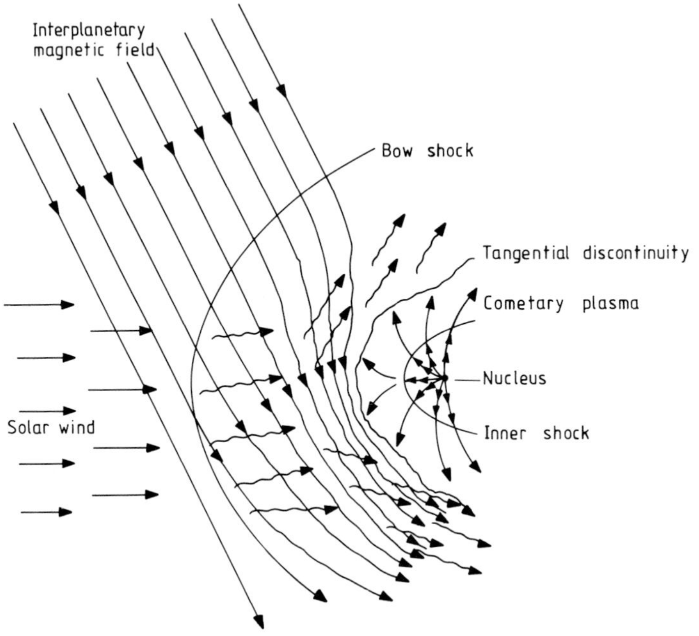

Figure 5.22 Interaction between the solar wind, interplanetary magnetic field and a comet (after Brandt and Chapman).

The interplanetary field has also been used to explain disconnection events in comets. A disconnection event occurs when a comet apparently discards its tail. The earliest such event to be observed was described in the 1890s by the astronomer Barnard. Such an event could occur when a comet crosses an interplanetary sector boundary (see figure 5.23). Reconnection of magnetic field lines cuts those lines anchored in the vicinity of the nucleus, and consequently strips away the magnetic field from the old sector to produce a completely disconnected plasma tail. A new tail is formed from the field lines in the new sector. This theory was first proposed to account for the disconnection event observed in Comet Kahoutek on 20–21 January 1974. An extensive search of the literature turned up 40 such observations and many of them can be correlated with the times of encounter between the comets concerned and sector boundaries. For modern events it is possible to correlate this directly with observations on the solar wind, while for older events this correlation is indirect and is based on the relationship between the speed of the solar wind and certain recorded geomagnetic indices.

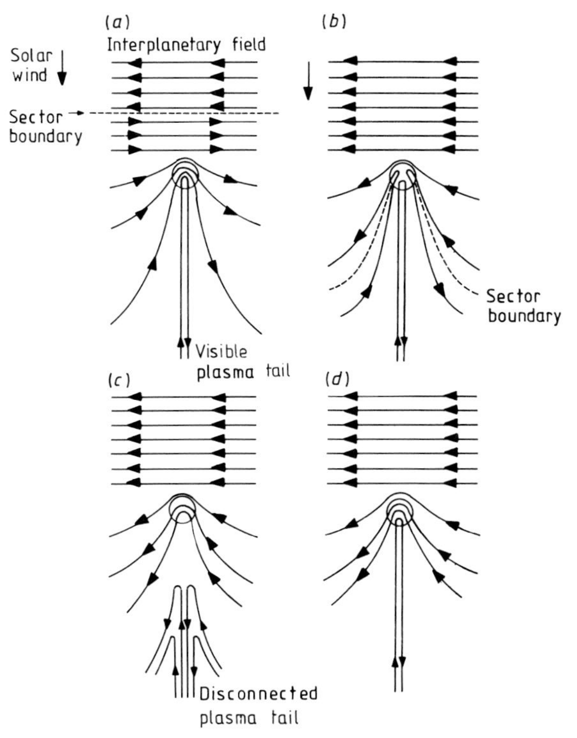

Figure 5.23 Sector boundary reconnection model of a cometary disconnection event (after Brandt and Chapman).

The Dynamo Problem

Now that we have discussed the magnetic fields in our solar system it is possible to make some general comments about the dynamos that may be operating in the various bodies that possess magnetic fields of internal origin. The dipolar component of a body's magnetic field can be quantified by a quantity called the magnetic moment, M, and its total angular momentum is usually denoted by L. If the magnetic moment of each body that possesses a magnetic field is expressed in terms of the magnetic moment of Earth, and if the angular momentum of each body is expressed in terms of that of Earth, then an approximate relationship exists between these quantities. This is often known as the magnetic Bode's law for solar system magnetism. It is expressed graphically in figure 5.24. Several people have argued that a fundamental relationship must exist between the two quantities M and L.

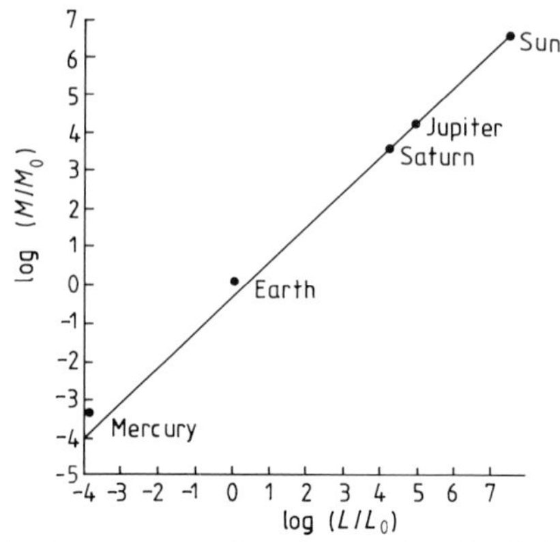

Figure 5.24 Graph of the magnetic Bode's law. (From Re-evaluating Bode's law of planetary magnetism by C T Russell *Nature* **272** 1978.)

Others have tried to produce scaling laws based on dynamos which are powered in the same way. However, no such scaling law is completely satisfactory and this may well point to the fact that the dynamos in the solar system may not all be operating in the same way. These studies have led to some suggested criteria for a dynamo to operate within a body. Although these conditions have not been rigorously proved, there seem to be three basic requirements for a dynamo to exist. There must be a sufficient volume of fluid; there must exist a sufficient source of energy; and there must be sufficient rotation. A major problem lies in what is meant by 'sufficient' in each case, and in trying to find quantitative criteria for describing the sufficiency. Investigations will, it is hoped, yield a better understanding of the dynamo problem.

Concluding Comments

The study of planetary magnetic fields is important for several reasons. It provides a testing ground for theories on the origin of planetary magnetic fields, and it can put constraints on models of planetary interiors. The interaction between the solar wind and planetary magnetospheres is helping scientists to understand some aspects of solar–terrestrial relationships. The magnetosphere of Jupiter is a unique plasma physics laboratory which can be used to test theories developed in other contexts. The Jovian magnetosphere and its radio emissions also serve as a model for the magnetospheres and radiation of pulsars. The study of magnetic fields in these objects, which may be seen as being on different rungs of the 'ladder' of cosmic distance, is necessary to enable study of the magnetic fields of extremely remote objects.

Chapter Six

Magnetic Fields in Stars and Pulsars

Since the Sun is a rather ordinary star of average size, and it possesses magnetic fields, it might be expected that most similar stars should also have magnetic fields. In this chapter we look at the astronomical evidence which shows that many Sun-like stars do have magnetic cycles similar to that of our own Sun and certain stellar classes have much stronger magnetic fields. The strength, detectability and behaviour of stellar magnetic fields varies from class to class, and these fields seem to be related to aspects of stellar evolution. We begin with a brief review of stellar classification and relevant aspects of the evolution of stars.

Classification of Stars

The dark absorption lines crossing the continuous spectrum of a star can be used to classify it. Originally the varying strengths of a particular set of lines in the hydrogen spectrum, the Balmer lines, were used to classify stars into classes which were labelled alphabetically from A to P: A stars had the strongest hydrogen lines. Soon after this system had been introduced, it was discovered that the spectral lines of the various elements and compounds have widely differing strengths in stars with different temperatures. As a result a new system was introduced which ordered the stars in a decreasing temperature sequence. This led to a rearranging of some of the alphabetically labelled classes and a dropping of other classes.

Most stars can be classified under one of seven spectral types, labelled by the letters O, B, A, F, G, K, M. Three further minority types have been added to this scheme—W stars at the upper end of the temperature sequence and R and N stars at the lower end. The order of this classification may be remembered by using the mnemonic 'WOw, Be A Fine Girl, Kiss Me Right Now'! The most important features of these classes are shown in figure 6.1.

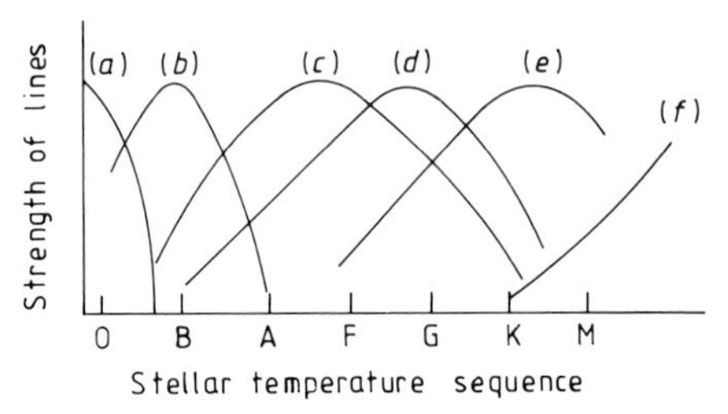

Figure 6.1 Graph showing the strength of different spectral lines in the spectrum of different stellar classes. (*a*) Ionised helium; (*b*) neutral helium; (*c*) hydrogen; (*d*) ionised metals; (*e*) neutral metals; (*f*) molecules.

These spectral types can be further subdivided into 10 subclasses denoted by numbers 0–9, placed after the spectral type letter. Special spectral characteristics within a subclass are indicated by an additional small letter placed after the spectral type.

Stars also vary a great deal in absolute brightness or, to use a more objectively measurable quantity, luminosity. The luminosity is defined as total energy radiated per second from a star. The Hertzsprung–Russell diagram (figure 6.2) shows the relationship between the spectral

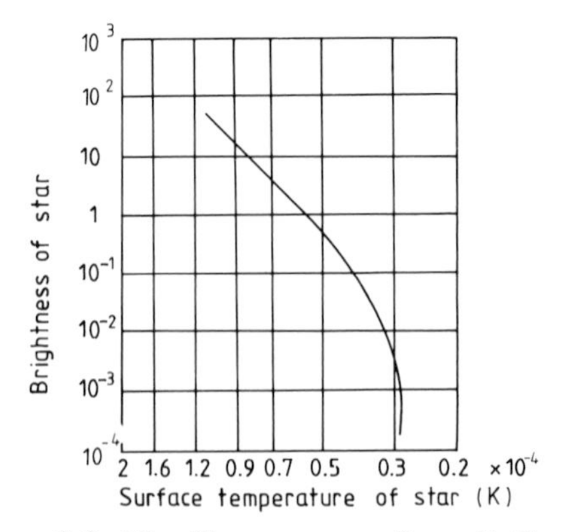

Figure 6.2 The Hertzsprung–Russell diagram.

type and the luminosity of stars. The brightest stars are also the hottest stars. The luminosity of a star is also related to its mass, more massive stars being more luminous than less massive stars.

The internal structure of a star is related to its total mass. For our Sun and stars of similar or lower masses, about 1% or more of the mass is in the form of convective envelopes. For stars between 15 and 1.5 solar masses, between 38% and 6% of the mass is in the form of convective cores. The importance of convection to the generation of magnetic fields by the dynamo mechanism has already been discussed. The fact that all main sequence stars have convective regions provides an important reason for believing that many stars should have magnetic fields at least as strong as that of our Sun.

Aspects of Stellar Evolution

Stars are formed out of the gravitational collapse of vast gas and dust clouds that exist between the stars. As such a cloud, many hundreds or thousands of times the mass of our Sun, contracts under the force of gravity it will usually fragment into smaller clouds with masses close to the average mass of a star. As these smaller clouds collapse even further the temperature, pressure and density of the centre of the cloud will increase considerably until the temperature is high enough for nuclear reactions to occur. At this stage the cloud becomes a main sequence star and most stars will spend the major part of their evolution in this phase. During this stage the star is fuelled by converting hydrogen into helium in the core.

Once all the hydrogen in the core has been converted into helium, the energy generation required to prevent the star from collapsing under the force of gravity ceases. At this stage the core collapses until sufficiently high temperatures are reached to burn hydrogen in a shell around the now inactive helium core. This is the shell burning phase. The outer envelope of the star expands considerably and cools as it does so. As a result the star will move off the main sequence to become a red giant. Although the outer envelope of the red giant expands considerably and becomes less dense, the core itself contracts a little and becomes more dense. As the density increases so the matter in the core assumes a new state called degenerate matter.

Degeneracy is a quantum mechanical effect, and in order to understand the concept it is necessary to explain two fundamental principles of quantum mechanics. The first is called the Pauli exclusion principle. According to this principle only two electrons, spinning about their own axes in opposite directions, are allowed to exist in a very small volume of special space—used to describe the state of the electron—called phased space. The size of the volume in phase space is determined by Heisenberg's uncertainty principle. According to this principle it is impossible to know both the position and the speed (or more precisely the momentum) of a particle with the same accuracy at a given time. In other words, if we know the position very accurately then we cannot know very much about its momentum. As a consequence of this law, if an electron is confined to a small region of ordinary space, its position will be known with a great deal of accuracy, but little will be known about its momentum, i.e. it will be moving about in all directions at varying speeds rather like a caged bird. As a result of this random motion the electron exerts a pressure on the walls of its container. This pressure is called degenerate pressure. It is this pressure that exists in the core of a red giant star.

In the red giant phase hydrogen burning continues in spherical shells further and further from the centre of the star, and the resultant helium is added to the core. The degenerate matter in the core contracts still further, and heats up until a temperature of one hundred million kelvin is reached. At this point helium burning begins, in which helium is converted to heavier elements like carbon. Successive nuclear fuels are used up until iron is reached. When all the nuclear fuels have been used there is then no more energy to support the star against the force of gravity. At this stage the star will contract considerably to become a white dwarf star.

Most of the matter in the core of a star that collapses to become a white dwarf is degenerate, with the electrons stripped from their nuclei and all the particles packed very tightly together, and so the internal pressure comes from the degeneracy of the electrons. The star will continue to contract until the forces due to gravity are balanced by the degeneracy pressure of the electrons. The radiation from a white dwarf cannot arise from nuclear reactions, but the initial high temperature at the core of the collapsing star that gave rise to the dwarf will slowly leak away through its gaseous atmosphere. Theoretical calculations on the structure of white dwarfs show that such stars cannot have a mass exceeding 1.44 solar masses. This does not mean that all stars above this limit cannot form white dwarfs. It is possible for a collapsing star to shed some of its outer shell and so, provided the matter in the core does not exceed 1.44 solar masses, the star may still end up as a white dwarf. If the amount of material in the core does exceed this value, then it is quite likely that still further collapse will occur and the star will end up as a neutron star or as a black hole.

Neutron stars are believed to be formed when the matter in the core of a collapsing star exceeds 1.44 solar masses (this limit is called the Chandrasekhar limit after the astronomer who first worked out the theory of white dwarf stars). If the core mass exceeds this value, then the force of gravity is greater than the degenerate electron pressure, and the resulting further collapse will crush the electrons into the nuclei to form neutrons. Neutron stars are prevented from further collapse by degenerate neutron pressure. However, if the mass of the collapsing core exceeds about three solar masses, then the force of gravity is even stronger, and the degenerate neutron pressure is not sufficient to support yet further collapse of the core to form a black hole.

A collapsing star becomes a black hole when its radius has shrunk below a certain limit known as the Schwarzschild radius. Below this limit the surface gravity is so high that no radiation originating in the sphere with this radius, and centred on the centre of the spherical mass, can escape from this region. This spherical surface is the 'event horizon' and it defines the boundary inside which all information is trapped. The spherical mass will itself still continue to contract since a black hole does not really have a steady state. However, the event horizon will be fixed with respect to the central mass.

Although the total mass of a star is obviously an important contributory factor in determining the final evolution of the star, the details are by no means clear. Theories that attempt to deal with steady mass loss from stars, catastrophic mass ejection and supernova explosions are still at a very early stage. However, it does seem possible to draw some general conclusions, although these conclusions are very tentative at the moment.

Stars with less than one solar mass have lifetimes that are longer than the present age of the universe. If the mass of a star is greater than or equal to 1 solar mass, but less than 1.44 solar masses, the star is very likely to end up as a white dwarf. If the mass of a star is greater than 1.44 solar masses, but less than 3 solar masses, it may end up as a neutron star or, by ejecting mass so that the remaining core is less than 1.44 solar masses, it may also end up as a white dwarf. For stars greater than 3 solar masses the situation is unclear. They could either end up as black holes, or by ejecting mass, as neutron stars. More theoretical research is necessary before these questions can be resolved.

Detecting Magnetic Fields of Stars

The great distances to other stars compared with the distance to our Sun presents a major difficulty in the investigation of stellar magnetic fields. In the chapter on the Sun it was seen that magnetic polarity fluctuated over the visible disc of the Sun. The field in sunspot pairs has different polarities in each half of the pair: the dipolar field which is most evident near the poles is naturally of opposite polarities, and the general field varies in polarity from one region to another. However, because of the closeness of the Sun it is possible, with even moderate-sized telescopes, to study the different parts of its visible surface separately. This cannot be done with other stars, where it is only possible to study the combined light of each star. The effects of different polarities will tend to cancel each other out, and no net effect will be measurable by the Zeeman splitting of spectral lines in Sun-like stars. The application of the Zeeman effect to the spectra of certain other classes of star has shown that some of them do possess magnetic fields which are much stronger and more ordered than that of the Sun.

Magnetic Ap Stars

Many of the strongly magnetic stars are a special subclass of A type stars, called A peculiar or Ap stars. The proportions of different chemical elements that exist in a given class are more or less the same. However, in the Ap stars certain elements such as manganese, silicon, strontium and the rare earths are more abundant than in normal A type stars. Other peculiarities of these stars are variations in their brightnesses and changes in their spectra.

Some Ap stars are rotating very rapidly, like many other kinds of star. Because of the Doppler effect it is difficult to detect the Zeeman splitting of lines in the spectra of rapidly rotating stars. As a star rotates part of the star is travelling towards Earth and part is travelling away. As we are studying the combined light from the whole star the spectral lines will be broadened by this rotation. It is this broadening that makes it difficult to detect the slight Zeeman splitting that may be present due to a magnetic field of the star. Astronomers have been able to measure magnetic fields in most Ap stars that are not rotating fast enough for the Doppler broadening to mask Zeeman splitting of spectral lines. These measurements show that the magnetic field being measured changes periodically, and the period of change is in most cases also reflected in other periodic changes in the spectra and light output of these stars. The observations can be understood in terms of a rather simple model. In this model we have a dipole field with an axis at right angles to the axis of rotation (see figure 6.3). The magnetic field near the poles of the dipole will affect the rate at which certain chemical elements diffuse upward to the surface of the star, so the concentration of chemical elements at the poles will be different from elsewhere on the star. This concentration will affect not only the observed spectrum of a star, but also its light output. Thus as the star rotates different parts of the star will point towards the observer, and the measured magnetic field, light output and spectrum will all vary with the same period.

The above model is not the only one that has been proposed. Another possibility is suggested, called the oblique rotator model, in which the magnetic dipole is not aligned along the rotation axis, but is inclined to it at an angle of less than 90°. Both models are able to explain the observations, and the question of which is the correct one depends on which theory for the origin of the magnetic field turns out to be the more acceptable.

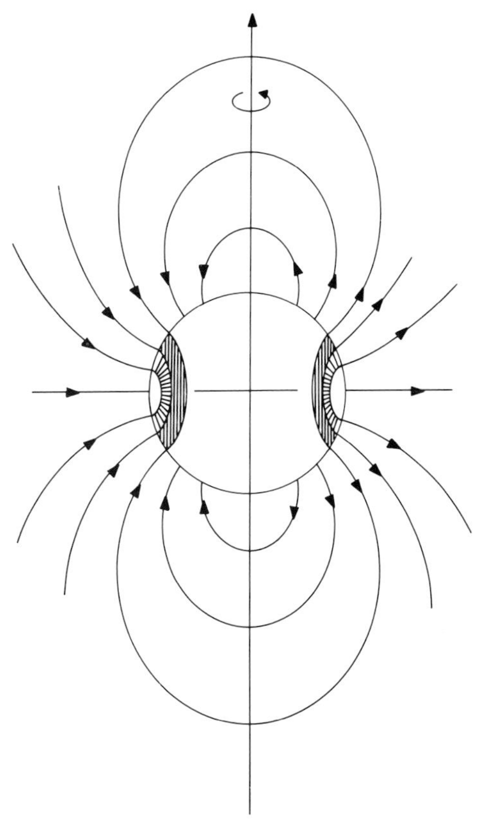

Figure 6.3 Model of an Ap magnetic star with dipolar axis at right angles to the rotation axis.

One possibility is that the field is generated by a dynamo action similar to that which is believed to operate in the planets and our Sun. Some scientists working on this theory have argued that the most stable arrangement for such a dynamo in stars with very strong fields is one with the dipole axis at right angles to the rotation axis. Another theory proposes that the magnetic field of the interstellar medium, out of which new stars are formed by gravitational collapse, was 'frozen' (or fossilised) into the star. The gravitational contraction of the proto-star would have compressed the field, thus making it much stronger than the original interstellar field.

Effect of the Interstellar Field on Star Formation

The interstellar field can also play a part in inhibiting star formation in the interstellar medium. Stars are born out of the gravitational collapse of interstellar gas clouds. This collapse is usually triggered by gravitational instabilities in the galactic gas. Magnetic fields in this gas can affect star formation in two separate ways. First, presence of a magnetic field alters the criteria which must be satisfied by fragments of the gas before they can start to collapse. Second, the actual collapse will be affected not only by the increase in gas pressure as the fragment contracts but also by the increase in magnetic pressure. Evidence for the existence of an interstellar field will be discussed in the next chapter.

Magnetic White Dwarfs

In the last few years magnetic fields have also been detected in white dwarfs. So far less than a dozen such stars have had their magnetic fields measured by the Zeeman effect, but several more are known to possess fields because of the polarisation properties of the radiation coming from them. The stars are believed to have dipolar fields, with strengths between 100 and 10 000 times more than that of magnetic Ap stars (see the Technical Appendix).

Magnetic Fields in Sun-like Stars

It has been known for some time that emission lines due to calcium are enhanced in the spectrum of those regions of the Sun, such as sunspots, where the magnetic field is strong (see figure 6.4). On the basis of these observations astronomers at Mount Wilson Observatory began a search for fluctuations in the calcium emission lines of a number of nearby Sun-like stars. Over a period of 14 years they detected such variations in a number of stars. From this they have concluded that some 91 stars similar to our Sun undergo cycles like the 22-year sunspot cycle of the Sun. The proportion of the stars they studied which turned out to be magnetic led these astronomers to argue that about half the stars similar to the Sun may have magnetic fields.

Since this work at Mount Wilson there have been several other investigations into stellar magnetic activity using different 'activity indicators'. From studies of our Sun it has long been obvious that solar magnetic activity is associated with x-ray and ultraviolet emissions from the corona. By analogy it would be expected that other Sun-like stars, for which it is difficult to measure magnetic fields directly, would

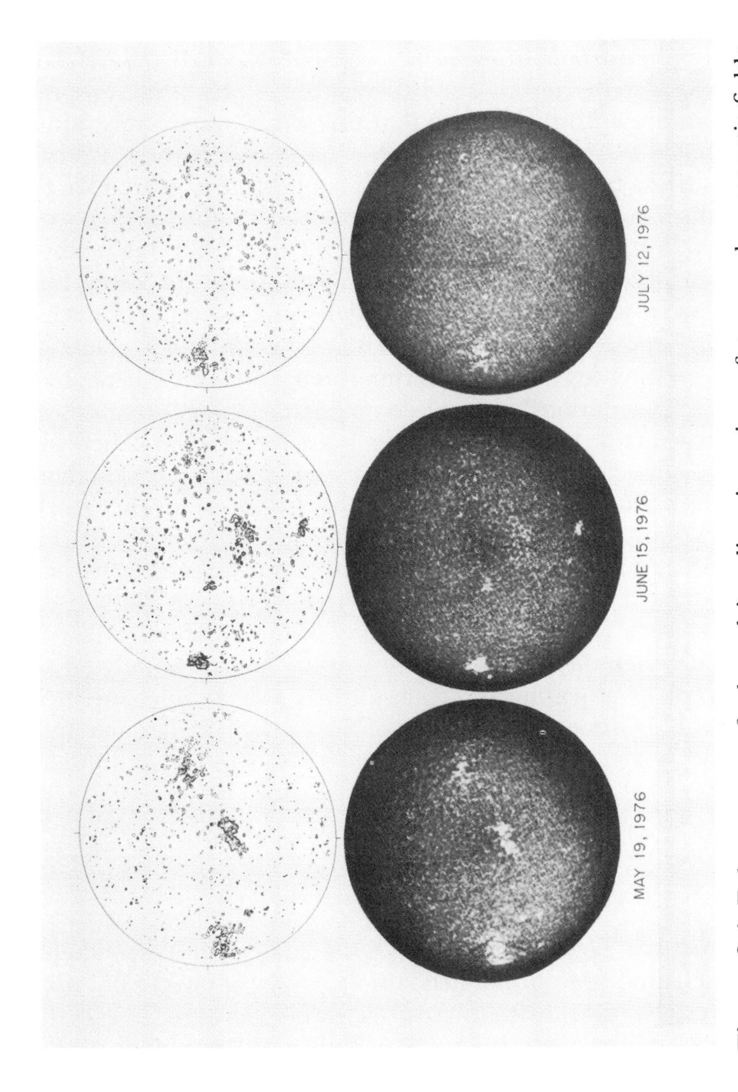

Figure 6.4 Enhancement of solar calcium lines in regions of strong solar magnetic fields. Magnetograms (top) and Ca II K-line spectroheliograms for the central area of the solar disc for the dates indicated. (Reproduced with permission from Large-scale solar magnetic fields by R Howard *Ann. Rev. Astron. Astrophys.* **15** © 1977 Annual Reviews Inc.)

nevertheless exhibit the associated x-ray and ultraviolet output. The International Ultra-Violet Explorer and the Einstein Observatory satellites have been used to study different types of stars to search for these effects, and the results have been spectacular, leading astronomers to believe that magnetic activity in certain stellar types is fairly widespread.

The total light outputs from certain types of stars called flare stars, show variations that can only be explained in terms of a large, cool 'sunspot' rotating with the star, so that the star is appreciably fainter when the spot is facing towards the telescope. Some of these stars are one thousandth as bright as the Sun, but frequently they exhibit flares a thousand times larger than the largest observed on the Sun. In these cases the whole star will be brightened by the flare, and this is why it is possible to observe these flares from a very great distance. Thus 'starspots' can be used to study stellar activity in these types of star.

From these researches a more coherent, though still incomplete, picture is beginning to emerge concerning stellar activity. There seems to be a very clear relationship between magnetic activity, rotation rate and stellar convection, although the details are not all clear. There also seems to be evidence for a relationship between stellar age and activity. At least some of the general principles that have been established can be explained in simple quantitative terms by a model proposed by Parker. He suggested that a regenerative magnetic dynamo in a convective star that is rotating will produce magnetic fields which will inevitably rise to the stellar surface. Turbulent convective surface motions will continually 'jostle' these fields and it is the 'jostling' of the emerging fields that leads to plasma heating and consequently to a chromosphere and corona. The most important outcome of the work on stellar activity is that it has led to much closer links between solar astronomy and stellar astronomy.

Pulsars

In 1968, Antony Hewish, together with some radioastronomical colleagues at Cambridge, announced the discovery of rapidly pulsating radio sources (pulsars) with incredibly stable periods. The actual discovery was really made by Jocelyn Bell, a postgraduate student working with Hewish, but the initial discovery was followed up with much more detailed investigation by the whole team. This rapid pulsation presented a severe problem for theorists, since it was difficult to explain the pulses in terms of the vibration of a white dwarf star. Although the existence of neutron stars had been proposed by two astronomers, Baade and Zwicky, in 1934, it was also difficult to see how neutron stars could pulsate with this period. Thomas Gold, a well known cosmologist and astronomer, suggested that pulsars were rotating neutron stars, but it was some time before this suggestion was developed into a theory able to explain the growing number of observations.

Observations of the first pulsar were made at radio wavelengths of a few metres, and these observations revealed a pulse period of about 1.3 s. Although the pulse rate was steady the amplitude varied from pulse to pulse. The arrival time for a given pulse varies with frequency, lower-frequency pulses arriving at a later time than their higher-frequency equivalents. This delay is due to the fact that the interstellar medium has a refractive index that is related to frequency, i.e. waves of different frequencies travel at different speeds. The delay can be used to estimate the distances to pulsars. This subject will be described more fully in the next chapter.

The pulses are all polarised, although the plane of polarisation is different at different times within a given pulse. The angle of polarisation is frequency dependent, showing that Faraday rotation is occurring in the interstellar medium.

Although at first pulsar periods seemed to be constant, further investigations showed that they increased slowly with time. The rate of increase of period with time has been measured for several pulsars. Another property of the period of some pulsars is that they occasionally take a sudden downward jump followed by the normal steady slowing-down pattern. These jumps have been called glitches, and they are believed to be starquakes due to sudden releases of strain in the crust of the neutron star. Just as earthquakes can be used to probe the interior of Earth, so glitches have been used to probe the internal structure of neutron stars.

The general structure of a neutron star is believed to be well represented by the model depicted in figure 6.5. A thin external crust consisting of a crystalline solid of nuclei, interlaced with degenerate high energy electrons, and about one kilometre thick, overlaps a thicker internal crust of about four kilometres. In this internal crust matter also has a crystalline form, but it is interlaced with electrons and neutrons. Below this is a neutron fluid, composed mainly of neutrons, but with some protons and electrons. The density of matter is so high in the solid core that it is difficult, given the present state of our knowledge about high densities, to theorise on the exact nature of this material.

Most of the pulsar's emitted radiation comes from its magnetosphere. It is now generally accepted that the neutron star at the centre of the pulsar has a dipole magnetic field which is not aligned with the rotation axis (see figure 6.6). Electrons spiralling in the magnetic field produce some of the radiation by the synchrotron effect. However, near the poles

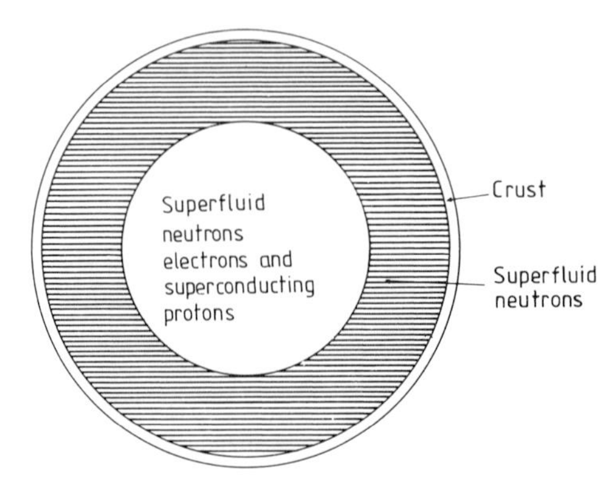

Figure 6.5 Cross section through a neutron star.

where the field is particularly strong, some of the electromagnetic waves arise from curvature radiation. In this process the helices of the electron orbits are so tight (i.e. the radii of curvature so small) that the electrons virtually follow the field lines. Since the magnetic field lines are generally curved any electron will be undergoing acceleration towards the centre of curvature and will radiate in the direction in which it is moving (just as for synchrotron radiation) because its speed is close to that of light. The radiation from either process will be particularly strong at the poles and this radiation will be beamed outwards. As the star rotates the beam will rapidly sweep across the radio telescope,

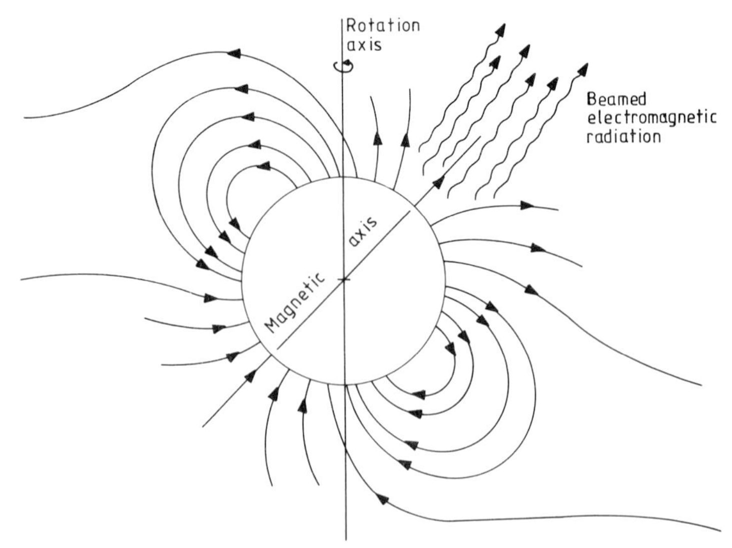

Figure 6.6 Magnetic field of a neutron star.

rather like the beam from a lighthouse, and this will be interpreted as a pulse of radio waves. Because of the intense magnetic field and the high energy of the electrons, pulsars would also be expected to pulse at optical wavelengths, and this phenomenon has been observed in the pulsars associated with the Crab and Vela supernova remnants. Some pulsars also pulse at x-ray frequencies, but this radiation is believed to be due to the accretion of matter onto the magnetic poles of the neutron star where the matter has been channelled by the magnetic field.

Only a small proportion of the emitted radiation is directed towards Earth, but even if this is taken into account, the total calculated loss of energy is insufficient to explain the rate at which the pulsars are slowing down. A much better approximation to the energy radiated is obtained by considering the pulsar simply as a rotating magnetic dipole. At any point in the region surrounding the dipole there will be a fluctuating magnetic field, and by the laws of electrodynamics there will also consequently be a changing electric field. From these fluctuations it is possible to calculate the rate at which the dipole is radiating energy and hence the rate at which it is slowing down, provided it is possible to calculate the strength of the magnetic dipole field. In other words the rate at which energy is being lost is related to the strength of the dipole, and hence the rate at which the pulsar's period is slowing down is a measure of the strength of the field. Pulsars with high magnetic fields will slow down more rapidly than pulsars with weak magnetic fields.

The slowing down of the pulsar is also related to the glitches. Any body with a solid crust will have its surface distorted, if it is spinning on its own axis, by the so-called centrifugal force. This force is proportional to the rate at which the body is spinning. In the case of planets and pulsars this spinning gives rise to an equatorial bulge. If the spin rate is changed the crust must change to accommodate the new equilibrium figure of the body. This means that the material in the crust is under stress as the rotation rate changes and it is this stress that can give rise to starquakes in the case of pulsars. It is these starquakes that give rise to the observed glitches in the periods of pulsars.

The pulsar in the Crab nebula has received special attention. The nebula itself has been the subject of intense astronomical research over a number of years, and it is a particularly interesting example of a supernova remnant. The explosion that gave rise to the Crab nebula was first seen by Chinese astronomers in 1054 AD. Over the last 40–50 years it has been studied by optical, radio, x-ray and theoretical astronomers, and at least some of the problems associated with it have been solved.

Radiation from the Crab nebula is highly polarised at both optical and radio wavelengths (figure 6.7). Astronomers now believe that almost all of its radiation is generated by the synchrotron process. The

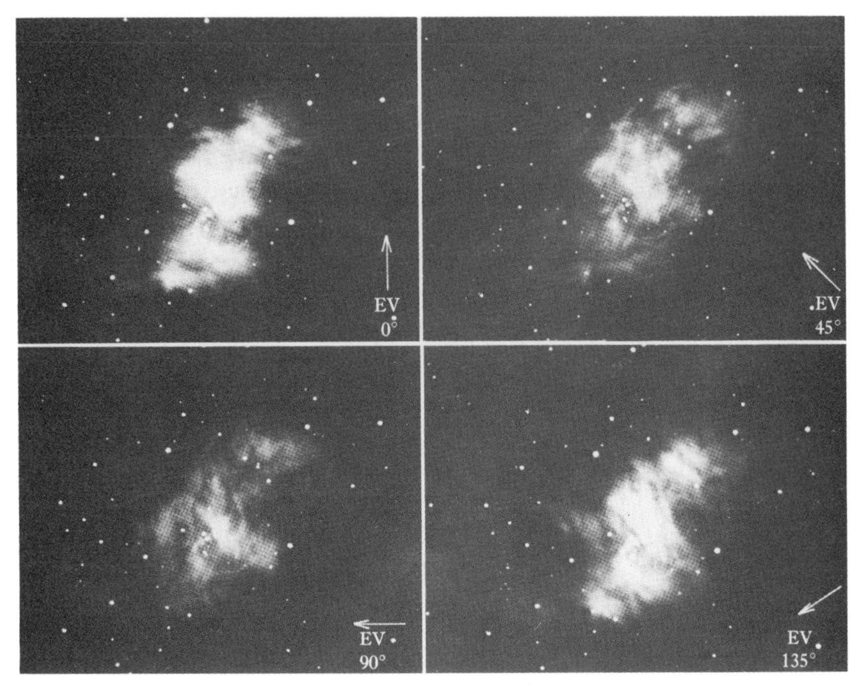

Figure 6.7 The Crab nebula observed in polarised light, with the electric vector at 0°, 45°, 90° and 135°. (Courtesy of the Hale Observatories.)

nebula is able to radiate by this process in such a wide range of wavelengths because the field is very strong, and the particles are very energetic. We have already seen that this type of radiation is polarised at right angles to the magnetic lines of force and so it can be used to give us information on the direction of the magnetic field in different parts of the nebula. All the energy required to produce this radiation is now believed to come from the pulsar near the centre. The rate at which the pulsar is slowing down is consistent with estimates of the rate at which energy is being fed into the nebula. This provides important evidence to support the theory outlined above.

Over the last few years radio astronomers have reported the discovery of a few pulsars with periods of a few milliseconds. What is also rather surprising about these discoveries is the very nearly constant rates of pulsation. On the basis of the generally accepted theory for the slowing down of pulsars, this would imply that such pulsars have a rather weak magnetic field. Some astronomers believe that these pulsars are very old and that the magnetic field of such objects does decay with time. One of these pulsars is a member of a binary system and its high rate of spin can be understood in terms of a large spin-up due to accretion of matter, and a consequent increase in angular momentum, from its companion.

Most astronomers at first assumed that the magnetic field of a neutron star was the remnant of the original star magnified by contraction. Recently, however, it has been suggested that thermal instabilities can occur in neutron stars and these could in principle support a dynamo mechanism. The available data on pulsars are really insufficient to decide at present which possibility is the more likely.

In 1974 a binary pulsar was discovered, and this system opened up the possibility of testing Einstein's general theory of relativity, so its progress was studied for several years. The presence of a companion was deduced from systematic drifts in pulse rate which could be attributed to the pulsar moving around another body with a period of some 7 hours 45 minutes. In other words the period of the pulsar was being Doppler shifted by its orbital motion. From the orbital velocity deduced by the application of the Doppler effect to the measured shift, it was possible to calculate the masses of the two stars. They both had a mass of about 1.4 solar masses and they are both thought to be neutron stars.

According to Einstein's general theory the point of closest approach of the two stars should precess with respect to other more distant stars. A careful study of changes in the orbits of the companions in the binary pulsar yielded results that are consistent with Einstein's prediction. This binary system also provided confirmation for another prediction of the general theory: according to the theory, a close binary system should lose energy via gravitational radiation and as a consequence the two members of the system should draw closer together and their orbital periods should decrease. This was in fact observed, and as a result we have the first observational evidence for the existence of gravitational waves.

Magnetic Fields in Double-Star Systems

Ever since their discovery, double-star (binary) systems have played an important part in astronomy. They are the only stars for which it is possible to determine their masses with any precision. It now seems likely that they also hold important clues to stellar magnetism. In some such systems, in particular the RS Canum Venaticorum, BY Draconis and some of the W UMa types, there are observed light variations which are different from the orbital periods. These could be explained in terms of the presence on one of the stars of large, cool regions (or starspots) rotating with one of the stars. This would seem to indicate the presence of magnetic fields.

For the RS Canum Venaticorum type the presence of magnetic fields is further supported by flaring, polarised radio emission, strong x-ray and ultraviolet emission. More direct evidence has recently come from spectroscopic studies of these systems. Certain spectral lines are more sensitive to Zeeman splitting than others. By comparing those lines sensitive to Zeeman splitting with those which are less sensitive it has been possible to infer fairly strong fields in at least one system. Studies of these star systems have thrown light on the possible relationship between stellar activity and age. In recent years it has become clear that age seems to affect stellar activity, mainly in that stars lose their angular momentum (i.e. their spin rate slows down) as they age. The RS Canum Venaticorum stars show high activity in spite of their age. This can be explained by attributing the maintenance of the rotation to the tidal coupling between the two components of the system.

Magnetic Fields in x-ray Sources

X-ray astronomers have discovered that generally the strongest point-like x-ray sources in our galaxy are binary systems in which mass is being transferred from one companion to the other. Magnetic fields also play a part in such systems. The x-ray source Hercules X-1 is fairly typical of some of these systems. This system consists of a neutron star with a spin period of 1.24 seconds which orbits a companion with a period of 1.7 days. This orbital motion leads to the neutron star being eclipsed by its companion but, because of the Doppler effect, it also leads to variations in the observed period of the neutron star. A diagrammatic version of the model proposed to explain this source is shown in figure 6.8. The neutron star Hercules X-1 is accreting matter from the normal star H2 Herculis. The incoming gas will be funnelled to the polar regions of the star's strong magnetic field (see figure 6.9). The energy gained by the gas as it falls towards the compact neutron star will cause it to heat up to a very high temperature: consequently the gas will emit x-rays. Some of the x-rays from the neutron star will be intercepted by the normal star, thereby causing an increase in the temperature of

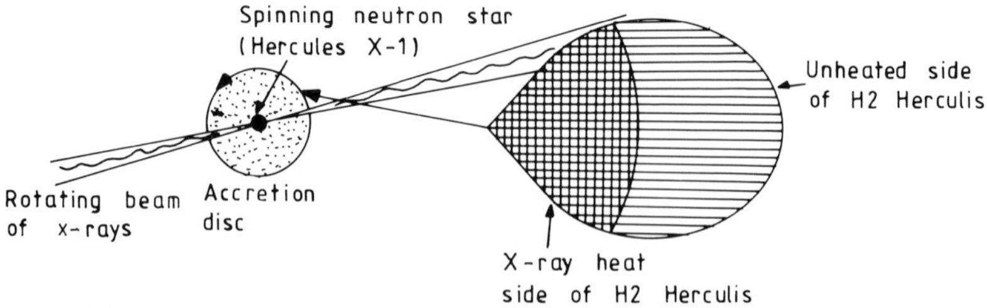

Figure 6.8 Model for the Hercules X-1/H2 Herculis system.

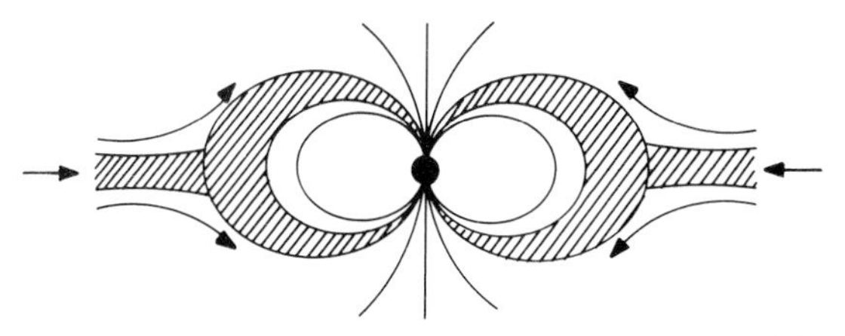

Figure 6.9 Accretion of matter by a neutron star with a strong magnetic field.

that side of H2 Herculis which points towards Hercules X-1. This increase in temperature is observable in the light waves received from the system.

Magnetic fields have also been invoked to explain another class of x-ray source known as the 'rapid burster'. If a neutron star is rapidly accreting matter, this matter will have a tendency to fall directly onto the star but the magnetic field will resist the tendency. A state of semi-equilibrium is reached in which the infalling matter is arrested in a standing shock wave caused by the magnetic field. However, an unstable situation is reached in which the build-up of matter can no longer be supported by the magnetic field and the sheer weight of the matter pushes aside the field lines, causing a burst of x-rays as it strikes the surface of the neutron star (see figure 6.10). The process can then be repeated as the shape of the magnetic field is restored and a new build up of matter occurs.

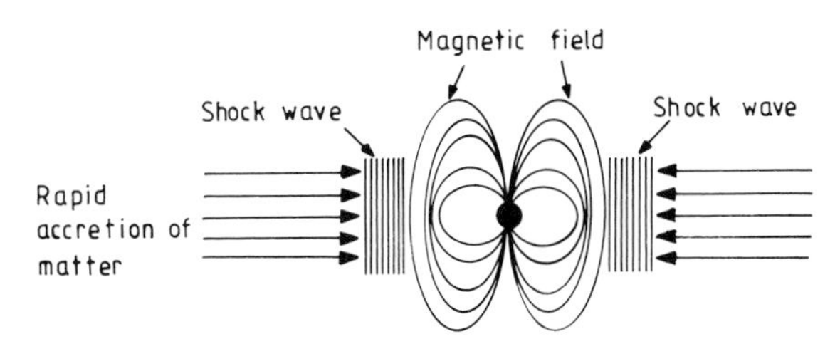

Figure 6.10 Model for an x-ray burster.

Concluding Comments

It is evident that magnetic fields play various roles in the activity, structure and dynamics of many kinds of star. In order to understand stellar structure and evolution it is necessary to know more about the nature and contribution of magnetic fields in stars and similar objects. Some of these objects provide the opportunity to study the behaviour of matter in very strong magnetic fields, and in this way research into these objects is complementing laboratory studies of magnetic fields. In order to understand the origin of these fields we must also know something about the interstellar magnetic field, which is described in the next chapter.

Chapter Seven

Magnetic Fields of the Milky Way

There is a vast magnetic field threading its way between the great distances that separate the enormous number of stars of our own Milky Way. In this chapter we look at how this field was discovered, the methods used to trace the lines of force, the part it plays in the structure of our galaxy, and its significance for our local Earthly environment. The magnetic field of the Milky Way galaxy is, however, related to its structure and dynamics, and some of the different galactic components are used to trace its structure, so this chapter begins with a view of the overall structure and dynamics of the Milky Way.

Structure and Dynamics of the Milky Way Galaxy

In many cultures the Milky Way is seen as a highway or road. For instance, the American Indian people, the Iroquois, see it as the 'road for souls' to the eternal kingdom; and the Hindu religion says it is the path to the celestial throne of Aryaman. Other peoples project aspects of their own homeland into their interpretations of the Milky Way. The Bushmen of the Kalahari desert think it is the reflection in the sky of the dying embers of campfires spread across the desert.

The Greek philosopher, Aristotle, also had ideas on the structure of the Milky Way. He thought the universe was divided into two separate regions, one below and one above the sphere of the Moon. Two different sets of physical law applied, he thought, in each region. Above the sphere of the Moon, the supra-lunar sphere, all was perfect and unchanging. In this region bodies were perfectly spherical and moved in perfect circles or circles upon circles. In the sub-lunar sphere, however, things tended to be imperfect, always subject to change, and bodies moved in non-perfect ways, in straight lines. Aristotle argued that, as the Milky Way was irregular, it could only belong to the sub-lunar region, and must therefore be an atmospheric phenomenon, and he included study of it in his work on meteorology.

Scientific study of the Milky Way really started with Galileo, an Italian astronomer and physicist who lived from 1564 to 1642. In 1609 Galileo heard that someone in Holland had invented a device consisting of lenses, which magnified distance objects. Within a few months he had made his own telescope and with it he discovered that the Milky Way really consisted of a large number of stars. William Herschel made the next great contribution to our understanding. He was born in Hanover in 1738 and started life in the German army. After coming to England he began to take a keen interest in astronomy. He taught himself to build telescopes and made several important discoveries including the discovery of the planet Uranus. Using one of his large telescopes he counted the number of stars of different brightness in different parts of the sky, and used this to work out the shape of the Milky Way. This was done by assuming that all the stars were equally bright, and therefore apparent differences in brightness were solely a result of stars' varying distances from Earth (figure 7.1). Since William Herschel's time the general size, structure and dynamics of the Milky Way has slowly been worked out, first with optical telescopes and more recently with radio telescopes. Let us now look at the present view of the Milky Way and the observations on which it is based.

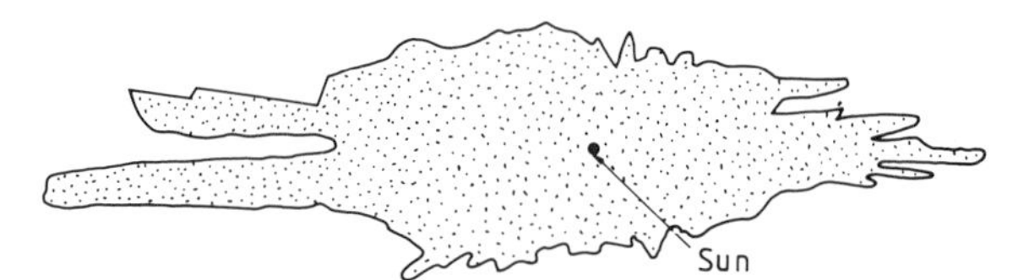

Figure 7.1 Herschel's view of the Milky Way.

A section through the galaxy at right angles to its plane shows it to be a collection of about 100 000 000 000 stars spread out roughly in the form of a disc with a bulge, called the nucleus, in the centre. Seen from above, a distinct spiral pattern would be evident for the distribution of the brighter stars (figure 7.2). All the stars are moving around the central bulge. Those near the plane of the Milky Way are moving in orbits that are very nearly circles, in much the same way as the planets move around the Sun. The stars are also moving at different speeds, those close to the centre moving faster than those further out.

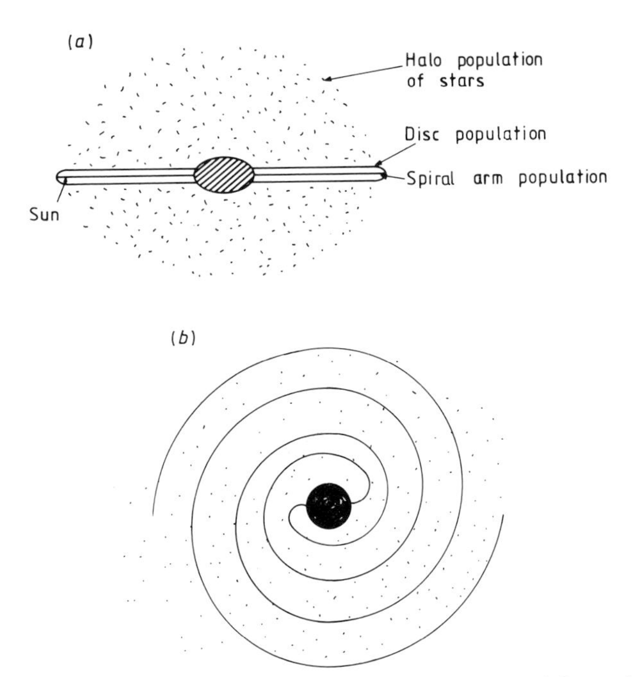

Figure 7.2 (*a*) A section through the Milky Way at right angles to its plane. (*b*) Spiral structure of the Milky Way.

Surrounding the whole galaxy is a halo of star clusters, with fairly large distances between neighbouring clusters. The orbits of these clusters around the nucleus are more elliptical than those in the plane of the Milky Way (figure 7.3). Two of the most important problems that had to be solved in working out the structure of our own galaxy were finding the distances to the stars, and deducing the size of the galaxy as a whole.

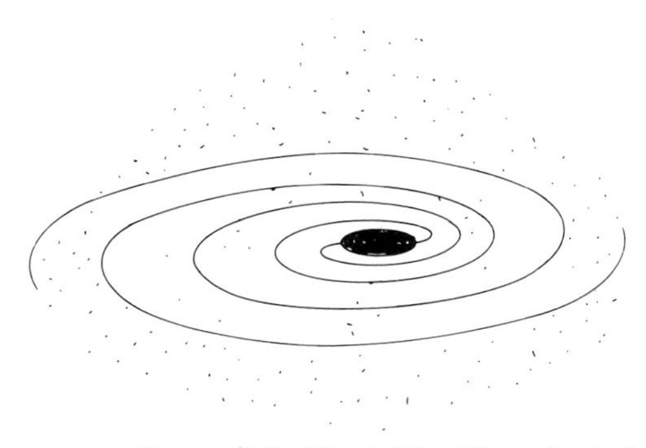

Figure 7.3 The Milky Way plus halo.

There are several ways of finding distances to stars. For stars that are closer than 150 light years, we can use the method of stellar parallax. This method is very similar to that used by surveyors for measuring distance on the surface of Earth. If a surveyor wants to measure the width of a river he can do so without crossing the river. He marks out a base line, then measures the two angles that an object, say a tree, on the opposite bank make with each end of the base line (figure 7.4). By drawing to scale the triangle formed by the tree and base line he can measure the distance to the tree on his drawing. The size of Earth is too small in comparison to distances of the stars to use this method in just this form. However, the apparent angular movement of a nearby star against the more distant stars, as observed from opposite points of the Earth's orbit, can be measured (figure 7.5). The diameter of Earth's orbit, rather than the size of the planet itself, is the effective base line in this case. This method becomes inaccurate for stars further than 150 light years from us, because the angular movement becomes so small. However, as astronomy progressed further, it was soon discovered that other properties of stars could be used to find distances to them.

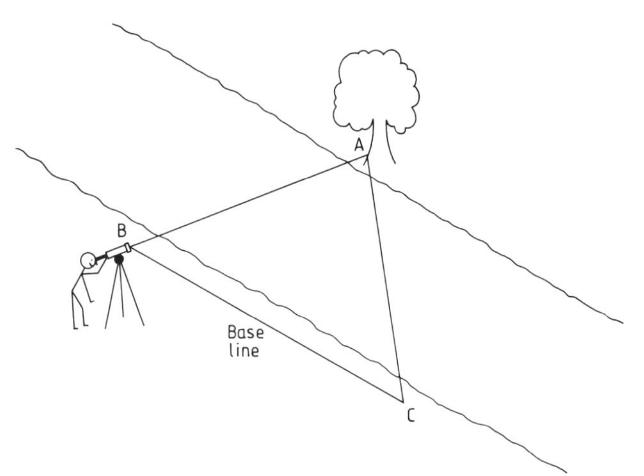

Figure 7.4 A surveyor measuring the width of a river.

The light we receive from a star depends not only on the star's intrinsic brightness, but also on its distance from Earth. If we know how much light a star is emitting, and use a special light meter to measure the amount of light received, then by comparing the two brightnesses we can work out the distance of the star from Earth. The number of watts marked on a light bulb is a measure of how much light it will emit, but how much light we receive from it depends on how far we are from it (figure 7.6). Do stars carry labels telling us how bright they are? Yes, they do.

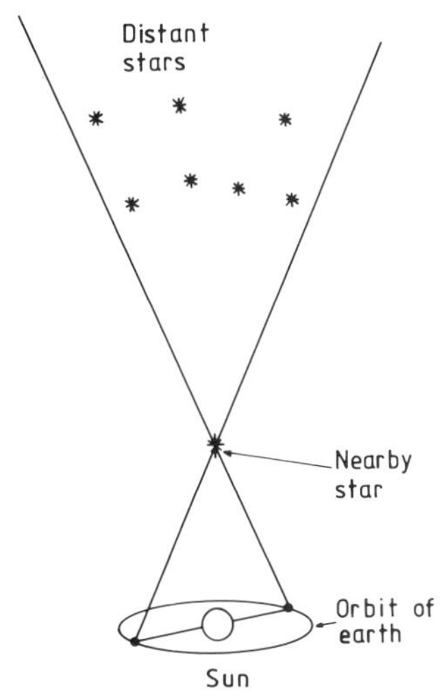

Figure 7.5 Nearby stars show parallax with respect to more distant stars as a result of the Earth's movement around the Sun.

In the last chapter we saw how stars are classified using their spectra, and we saw that the spectrum for a given star was related to its temperature, and for all stars this was related to its actual brightness. The spectrum of a star is therefore a label which gives us a good idea of how much light it is emitting. The brightness of a star as seen from the surface of Earth can be measured, and if we compare this brightness measurement with the actual brightness indicated by its spectrum, we can work out its distance from us.

There are other labels carried by stars which give clues to actual brightness. One such is associated with a special type of star known as a

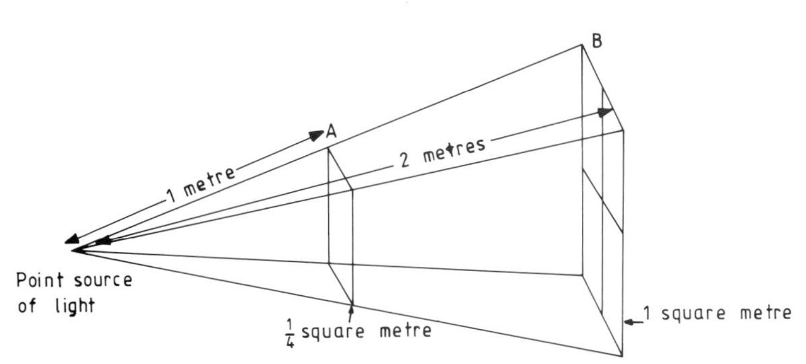

Figure 7.6 The inverse square law. The light passing through ¼ square metre at A passes through 1 square metre at B. As a result the light energy per unit area at B is one quarter of that at A.

Cepheid variable. The light emitted by a star of this type varies with periods ranging from fractions of a day to several days. The period of variation is related to the average actual brightness of the star (figure 7.7). By comparing this with the brightness which we measure on Earth we can find its distance. This method works for Cepheid variables in our own and other galaxies.

Although the method described above can give us some idea of the distances to stars in our own Milky Way, we cannot use it to find the size of our galaxy because there are tiny dust particles in the plane of the galaxy. These particles obscure the light from the more distant parts of the Milky Way.

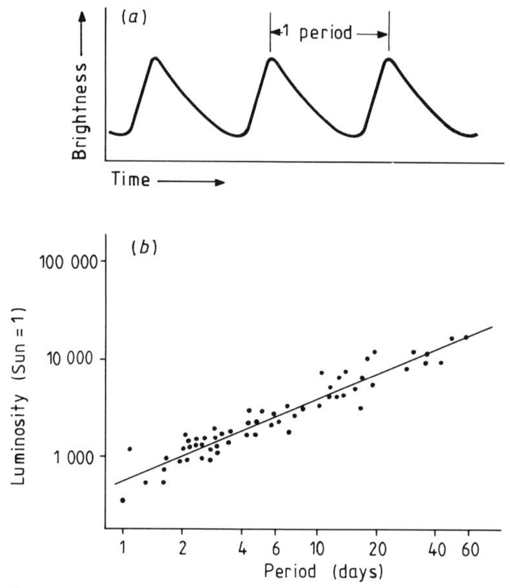

Figure 7.7 (*a*) Variation of the brightness with time of a Cepheid variable star. (*b*) The relationship of brightness with period for Cepheid variable stars.

Earlier in this chapter we described the 'halo' of the galaxy as a spherical distribution of star clusters which surround the whole Milky Way. The globular clusters of stars in this halo were used by an American astronomer called Shapley to estimate the size of the galaxy. He used three different methods to do this. First, some of the clusters contained Cepheid variables, so he could use the method suitable for these stars, as described above, to measure their distances. Second, he made several assumptions, one of which was that the brighter stars in each cluster all had the same actual brightness, and so by comparing the brighter stars in a cluster of known distance with those in one of unknown distance he was able to say how far away the latter cluster was. Third, he assumed that all clusters were of the same actual size, and that the observed differences in angular size as seen from Earth arose purely from the differences in their distances from Earth. These last two methods may seem rather crude, but the three methods together gave Shapley a very good estimate of the size of our galaxy.

The methods used by Shapley worked because he was using clusters of stars that were out of the plane of the galaxy. The small dust particles mentioned earlier act like a fog, cutting down visibility in the plane of the Milky Way and making it difficult to measure distances accurately. However, this fog has little effect when looking out of the plane of the galaxy.

Unlike the planets, all stars, even those nearest to us, are so far away that they will only seem to change their relative positions in the night sky very slightly, even over a period of several years, and even though they are moving at very high speeds. This means that astronomers have to use special methods to study the movements of the stars, as distinct from finding their distances from Earth. For some of the nearer stars the movements can be measured by taking two sets of photographs of different parts of the sky, several years apart. When the two sets are compared small drifts in the positions of some stars will be apparent. The distances to several of these stars are known, so we can then calculate the speeds with which they are travelling across the direction in which we are looking (figure 7.8). However, this method does not tell us if the stars are moving away from or towards us. To find out this additional piece of information we must make use of the Doppler effect.

In the last chapter we saw that the dark absorption lines crossing the bright continuous spectrum of a star can be used to identify the elements

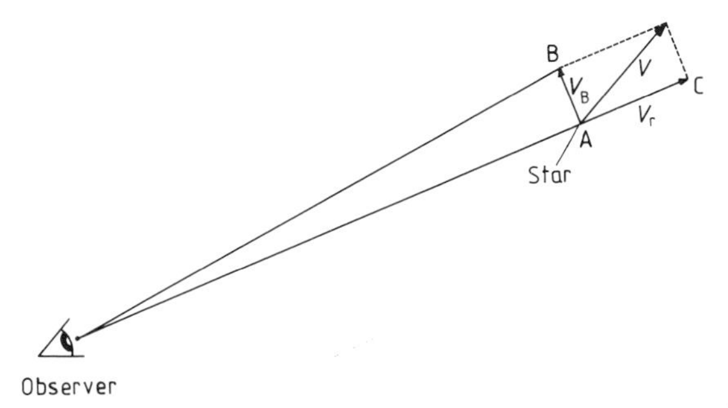

Figure 7.8 The star has a real velocity V in the direction shown. The observer can only measure directly the apparent change in position from A to B. The velocity V_r can be measured using the Doppler effect.

in the atmosphere of the star. However, if the star is travelling towards us these lines will be shifted towards the blue end of the spectrum, and if the star is moving away from us the lines will be shifted towards the red end of the spectrum. The amount of the shift towards the blue or red end tells us how fast the star is travelling and whether towards us or away from us.

The two ways of measuring movements of stars described above have allowed astronomers to make important discoveries about the motions of stars in the galaxy. The motion of each star in the galaxy is controlled by the force of gravitation between it and all the other stars in the galaxy. Just as we can find the mass of the Earth by measuring the force of gravitation which Earth exerts on a particle, so we can measure the mass of the Milky Way by measuring the force which it exerts on the individual stars. This in turn can be found by studying the motions of the stars.

Optical astronomy has given us a great deal of information on the size of our galaxy, the distances to stars, and their motions. Radio astronomy has increased our knowledge about what is between the stars. Far from being a total vacuum, there is a great deal of material in interstellar space—atoms of gas, subatomic particles such as electrons (which move about and which emit radio waves), and dust particles revealed by optical studies. The radio waves can be used to detect different types of atom or molecule, and to map the motions and distribution of these particles.

Hydrogen is the most abundant element in the interstellar medium, but several other atoms and molecules have also been detected. Using the Doppler effect, radio astronomers have found gas to be moving about the centre of the galaxy in much the same way as the stars. Just as radio waves are not affected by fog on Earth, radio waves from the galaxy are not affected by the dust particles which obscure light rays. Radio telescopes can therefore study the whole galaxy, and our picture of the Milky Way is much more complete as a result of this work.

Cosmic Rays and Magnetic Fields

The suggestion that the Milky Way galaxy may have a magnetic field was first made by the American scientists Fermi and Chandrasekhar. They were trying to explain the existence of cosmic ray particles which were coming to Earth from beyond the solar system. They suggested that these particles were ejected in violent supernova explosions. However, if this were so, the particles would be expected to come largely from the plane of the Milky Way where more supernova explosions are likely to occur. Instead they came from all directions in space. Another problem posed by this suggestion concerned the total number of particles observed. If these particles were produced in supernova explosions many would be ejected out of the galaxy, and only a small proportion would reach Earth. Both these difficulties could be resolved by supposing that a large-scale magnetic field existed in our Milky Way. The cosmic-ray particles are confined to the galaxy, despite their high energies, by the interstellar magnetic field lines around which they will spiral. These field lines are frozen into the thermal gas of the interstellar medium which is concentrated towards the plane of the galaxy by the gravitational field of the whole Milky Way. The cosmic-ray particles would, because of the field lines, come to Earth from all directions because they would be spiralling around the lines rather than moving along them (see figure 7.9). Much later on it was pointed out by Parker that such a situation was very unstable. The gas would tend to move towards the plane of the galaxy where the force of gravity is much stronger. The cosmic-ray particles are hardly affected by gravity and because of their energy would tend to move upwards. The magnetic pressure between the lines of force would tend to push the lines upward. However, because the field lines are embedded in the gas the situation cannot simply be inverted. The gas can move along the field lines and the magnetic field can be distorted upwards in between (figure 7.10).

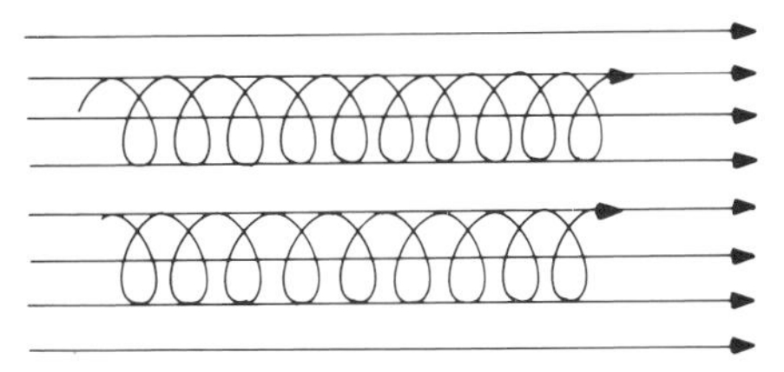

Figure 7.9 Containment of cosmic rays by the magnetic field of the Milky Way.

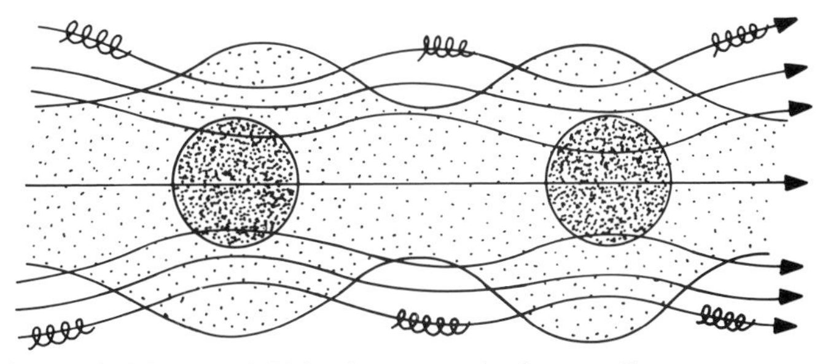

Figure 7.10 Instabilities between the interstellar gas, magnetic field and cosmic-ray particles.

Magnetic Fields and Spiral Arms

Soon after Chandrasekhar and Fermi first made their suggestion about magnetic fields and cosmic-ray particles, they also invoked the presence of the field to explain the spiral structure of the Milky Way. These scientists assumed that the bright young stars of the galaxy which were found in the spiral arms were there because they were still close to their places of birth in the interstellar gas, which has a spiral structure. Yet if the gas were in the form of spiral tubes, there was not enough gas pressure to support such a tube against the gravitational force of the galaxy. However, if the galactic magnetic field were in the form of spirals directed along the spiral arms, then there would be an additional magnetic pressure at right angles to the lines of force and this could supply the extra pressure needed to support the tubes of the spiral arms (figure 7.11).

Several years later Hoyle and Ireland showed that such support could be provided by a field in the form of helical lines of force wrapped around the axis of the spiral arms (figure 7.12). The modern theory of spiral structure is due to Lin and Shu. They suggested and developed a theory which in effect says that spiral structure is a density wave maintained by the self-gravity of the large-scale distribution of matter. As a result of this work magnetic fields are no longer considered important to spiral dynamics. However, these ideas did stimulate several other areas of research in galactic astronomy.

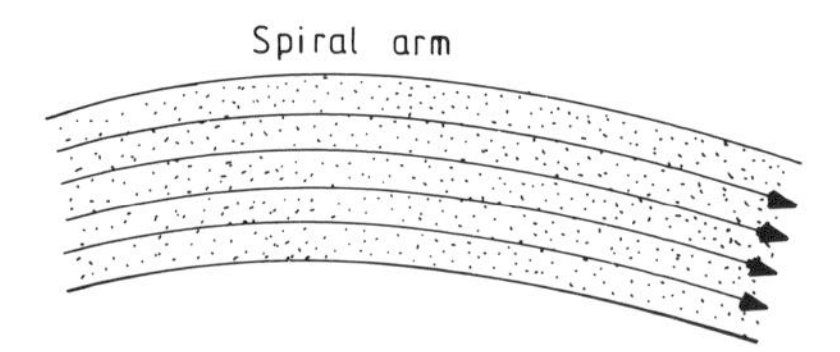

Figure 7.11 Magnetic field lines directed along spiral arms.

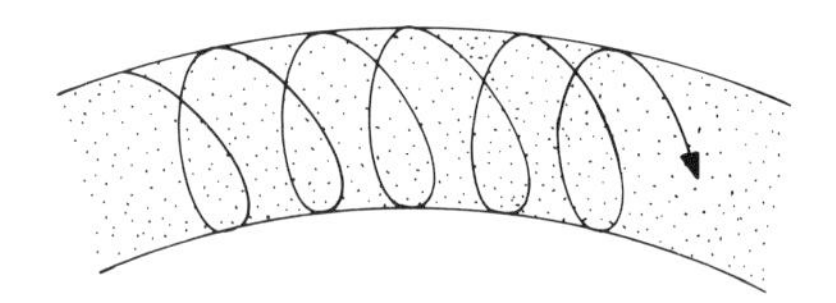

Figure 7.12 Magnetic field lines in the form of helices with axis along spiral arms.

Polarisation of Starlight

In 1948 Hall and Hiltner, two astronomers working at the USA Naval Observatory in Washington, built a polarimeter to measure the linear polarisation of stars. Their reason for starting a programme to measure stellar polarisation was a theoretical prediction by Chandrasekhar that under certain circumstances the light from eclipsing binary stars would be polarised. Chandrasekhar and Breen had suggested that the scattering of light by electrons was the chief source of opacity in early-type stars. Their calculations showed that a percentage of the light coming from the limb of such a star would be plane polarised parallel to the limb. They suggested that if an early-type star was partially eclipsed by another star, in a binary system, then it would be possible to detect this effect (see figure 7.13).

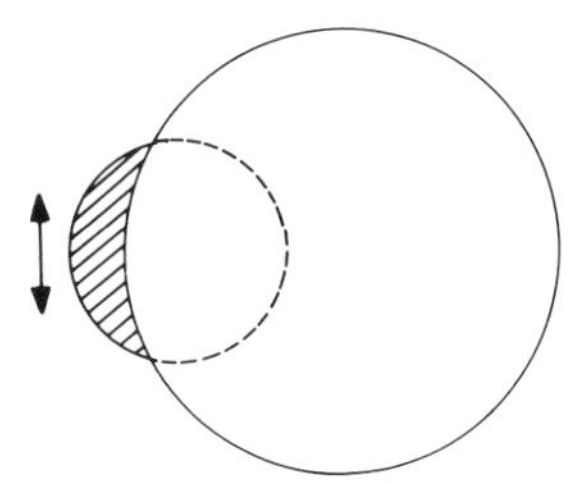

Figure 7.13 Polarisation of light from certain binary systems.

Hall and Hiltner set out to do just this, but instead they found that a large number of stars, most of which were not binary, were polarised. Their studies showed that the polarisation was not in any way related to spectral type. However, for stars in certain directions within the galaxy the amount of polarisation was related to the amount of reddening of the light from the star (due to absorption by interstellar dust). This suggested that the polarisation might be due to dust grains aligned by the interstellar magnetic field.

Modern theories of interstellar polarisation are based on the work of Davies and Greenstein. It is now generally accepted that the polarisation is due to scattering and absorption by plate-like dust particles. The magnetic properties of the particles are such that they spin around the lines of force, rather like a wheel on an axle. A beam of unpolarised light coming from a star will interact with these particles. The waves vibrating at right angles to the field lines will 'see' more of the dust particles and more of these waves will be absorbed. As a result of this interaction the light passing through the interstellar dust will be slightly more strongly polarised parallel to the magnetic field (figure 7.14). This theory predicts that the polarisation of starlight would be a maximum when looking at right angles to the field lines and almost zero when looking along field lines.

The polarisation of a large number of stars has now been measured and an analysis of the data seems to show that the magnetic field of the Milky Way is parallel to the plane of the galaxy. In the neighbourhood of the Sun the observations are consistent with a field pointing in a direction making an angle of 50° with the galactic centre (see figure 7.15).

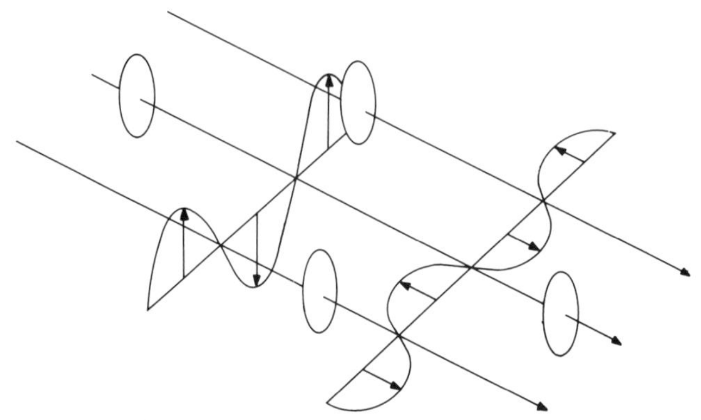

Figure 7.14 Interstellar polarisation by dust grains.

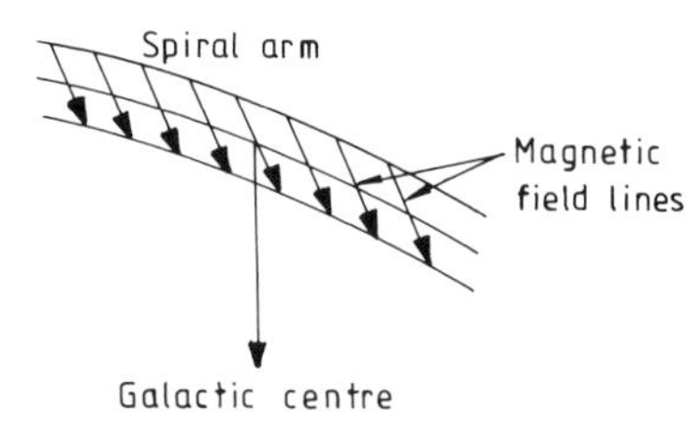

Figure 7.15 Magnetic field lines make an angle with the spiral arm axis.

Like all methods used in astronomy, where we do not have direct access to the objects which we are studying, this method of investigating the field has its advantages and disadvantages. An important advantage is that we can put distance limits on where the polarisation is occurring since this must be somewhere between Earth and the star being measured, but not beyond the star. This fact has been used by some people to put distance limits on features in the radio sky which are believed to be associated with anomalies in the magnetic field of the galaxy. One disadvantage is that the method can only be used to investigate the field in those regions where dust exists. This means that it is confined to the thin layer of dust near the plane of the Milky Way. Since the dust also reduces visibility in the galactic plane the method is only applicable to the region in which the light from the stars has not been obscured by absorption.

Synchrotron Radiation from the Galaxy

The fact that not all radio radiation from the galaxy could be accounted for in terms of thermal emission from a gas was already evident soon after the first radio map of the Milky Way was published by Jansky in

1932. However, it took several more years before it was realised that part of the radiation was due to cosmic-ray electrons spiralling in the magnetic field of the Milky Way. In 1950 Alfven and Herlofson proposed that the non-thermal part of the emission was produced by the synchrotron process when cosmic-ray electrons interacted with the trapping fields or 'magnetospheres' of stars. This model was no doubt inspired by Alfven's picture of the solar system where cosmic rays originated in the Sun and were made isotropic by an interplanetary magnetic field. It was left to Kiepenheuer to make the suggestion that this radiation was due to electrons in the galactic magnetic field. The application of synchrotron theory to astrophysical problems was further developed almost exclusively in Russia by Ginzburg and Shklovsky.

Some attempts were made to deduce the structure and strength of the galactic magnetic field from the observed distribution of the synchrotron part of the galactic radio radiation. These attempts were only partially successful, partly because it was difficult to separate the total from the synchrotron emission in any analysis, but partly because it was necessary to make several assumptions about the ratio of the ordered part of the field to the random fluctuations. Once the polarisation of this radiation had been detected it not only confirmed that the radiation was indeed due to the synchrotron process but also allowed astronomers to make some deductions concerning the structure of the local field.

Mapping the structure of the galactic magnetic field from the polarisation of synchrotron radiation is beset by practical difficulties. When a measurement is made in a given direction we are in fact detecting all the radiation within the beam of the telescope. If, as is now believed, the galactic magnetic field has small-scale fluctuations superimposed on a more ordered field, then the polarisation from the various parts of the interstellar medium within the beam will be orientated in different directions, and some of them will tend to cancel each other out. The observations are still further complicated by Faraday rotation if the field has a component along the line of sight. This is because the polarisation of elements at different distances from the telescope will be Faraday rotated by different amounts, and the net result will mean a decrease in the percentage of polarisation. However, despite these difficulties it is possible to make some deductions about the local field, where the effects just mentioned will be less important. When looking at right angles to field lines there will be no Faraday rotation and the radio polarisation should be at right angles to the optical polarisation in this region. This has in fact been observed near the plane of the galaxy in the direction making an angle of 140° with the galactic centre. This gives a direction for the local field which is consistent with that obtained using interstellar polarisation. This method gives information on the galactic field in those regions where there is an abundance of cosmic-ray electrons.

Faraday Rotation of Extragalactic Radio Sources

Radioastronomical observations of Faraday rotation in the radiation from extragalactic radio sources have proved to be one of the most effective ways of investigating the large-scale structure of the galactic magnetic field. The possibility that Faraday rotation may be occurring somewhere in open space was first suggested by Cooper and Price in 1962 when they noticed that the angle of polarisation of the radiation from Centaurus A varied with wavelength. As the number of extragalactic sources with measured rotation measures increased, so some researchers used the data to argue that the field pointed in opposite directions above and below the plane. This type of conclusion is only possible with Faraday rotation measures since the other methods discussed earlier on can give information on direction but not on the sign of the field, i.e. they cannot be used to tell if the field is pointing towards or away from the observer. Some workers noticed a correlation between the reversals of the field and the motion of neutral hydrogen gas. On the basis of this relationship they suggested that the reversals were due to the longitudinal field being drawn out by the flow of hydrogen.

After this early work there arose support from different quarters for three models of the magnetic field. The first one consisted of a loop of field superimposed on a longitudinal field directed along the local spiral arm. The second consisted of a longitudinal field that reversed on going from above to below the plane. The third was a sheared helical field wrapped around the axis of the local spiral arm. Figures 7.16, 7.17 and 7.18 illustrate these models. Since then the number of data have increased considerably, so the interpretation can be approached with more certainty, but there is still no general consensus of opinion.

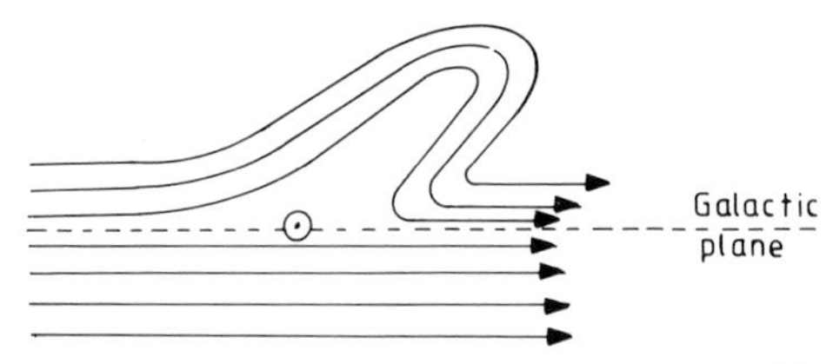

Figure 7.16 Longitudinal field along spiral arm with a loop near the Sun.

Recently two astronomers in Canada, Simard-Normandin and Kronberg, proposed a bisymmetric spiral field model, in which the field lines followed the spiral arms but reversed when going from one spiral arm to the next. This particular model has been supported by two

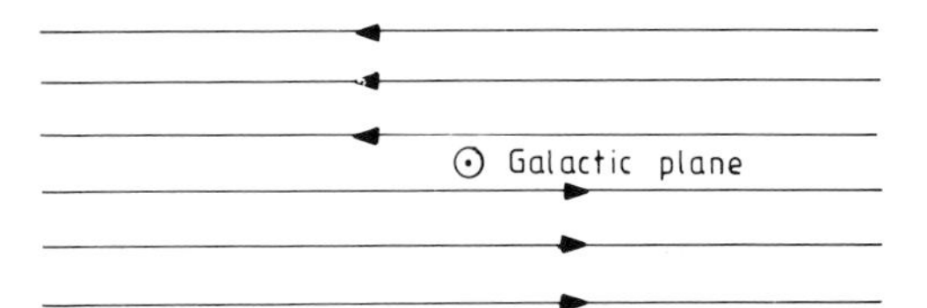

Figure 7.17 Longitudinal field along spiral arm, but in different directions on either side of the plane.

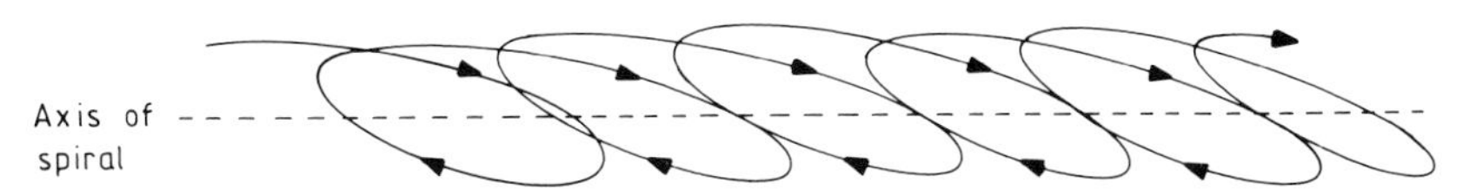

Figure 7.18 A sheared helical field with axis along the spiral arm.

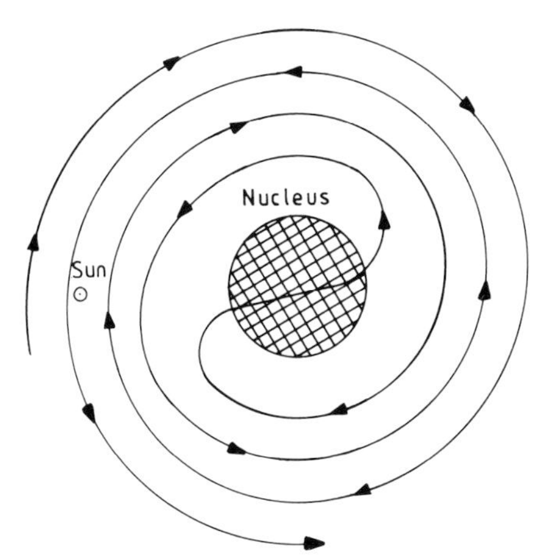

Figure 7.19 Bisymmetric spiral arm field.

Japanese astronomers, Sofue and Fujimoto (see figure 7.19). These two argue that the model is consistent with the hypothesis of a primordial origin of the field. If the whole universe possessed a magnetic field, the field would have been frozen into the gaseous material out of which the galaxies formed. The gravitational collapse of this material to form a galaxy would have led to an amplifying of the field and the winding up of the field to follow the spiral arms would have resulted from differential rotation. However, the age of the galaxy is such that, if this explanation were correct, there should be several more spiral arms than have been observed and they should be more tightly wound.

Pulsars and the Interstellar Medium

Pulsars have also been used to give information on the properties of the interstellar medium. Although all types of electromagnetic radiation travel with the same speed in free space, they do not travel with the same speed when gases or other transparent forms of matter are present in space. The pulses from pulsars can be studied at a variety of wavelengths by 'tuning' the radio telescope to different wavelengths. These observations show that the arrival times of the pulses at different wavelengths differ by small but measurable intervals. From these intervals it is possible to work out the speed of different waves in the interstellar medium. This gives some information on the electron density of space. Alternatively if an average electron density is assumed, it is possible to make estimates of the distances to the pulsars.

The radiation from pulsars is also linearly polarised and the angle of polarisation varies with wavelength. This is of course due to the Faraday effect and hence the Faraday rotation can be used to map the galactic magnetic field in much the same way as the rotation measures from extragalactic radio sources are used. All observations of rotation measures give information on the field in those regions where thermal electrons are present. However, there are fewer pulsars than extragalactic radio sources, and they are confined largely to the galactic plane, so the magnetic field information obtained from these observations is more limited. Nevertheless two astronomers in Wales, Thomson and Nelson, have used pulsar data to conclude that the galactic magnetic field exhibits a reversal in direction towards the inner spiral arm. They take this to provide support for the bisymmetric spiral arm model.

A Unified Model of the Large-Scale Galactic Magnetic Field

Using information from all available sources, there is general support for a model in which the field lines are concentric circles, about the centre, and in the galactic plane. However, these circles have been disturbed by shock waves associated with the formation of spiral arms. Such a field is consistent with a galactic dynamo, and it overcomes the difficulty of the winding-up of a primordial field as suggested earlier.

Evidence to support the bisymmetric spiral is not convincing. Some authors have argued that reversals in rotation measures for extragalactic radio sources are consistent with such a model, but rotation measures average the field over the whole line-of-sight area between the source and the observer and there is no positive way of knowing exactly where the reversal is occurring. It could well be due to local irregularities. The pulsar data are rather too sparse to support such a conclusion.

The most acceptable model for the large-scale structure of the galactic magnetic field is that shown in figure 7.20. Superimposed on this large-scale structure are local irregularities associated with such features in the radio sky as the North Galactic radio spur and the Cygnus loop. Both these features could well be supernova remnants and would thus represent regions in which the general field has been distorted by supernova explosions. There is also indirect evidence, from the percentage of optical and radio polarisation, that the field lines are further distorted on a very small scale by wavelike 'ripples' superimposed on the large-scale features already described.

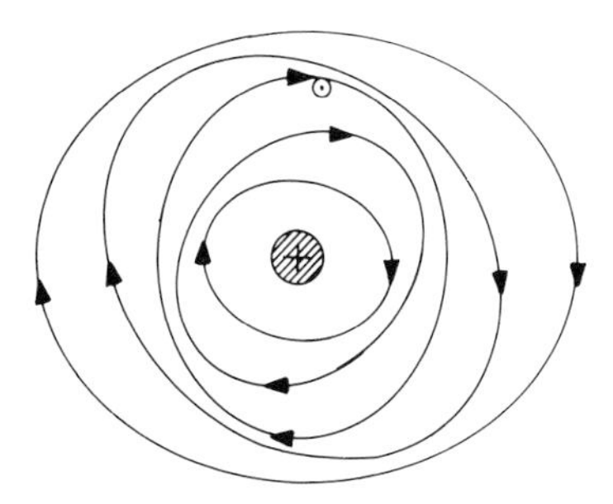

Figure 7.20 Field in the form of concentric circles distorted by the density wave pattern.

All the methods so far discussed can be used to make indirect estimates of the strength of the galactic magnetic field. The Zeeman effect has been used to make a direct measurement of the field strength. Atoms of neutral hydrogen gas in the interstellar medium emit radio radiation at a wavelength of 21 cm. However, in the presence of the interstellar field this line would be split into two circularly polarised components. The field strength obtained using this method is in general agreement with that obtained using other methods.

Summary and Concluding Comments

The chapter started with a brief overall view of the structure and dynamics of the Milky Way, because these aspects of galactic structure are intimately bound up with the interstellar magnetic field. In our galaxy the stars are kept together by the force of gravity between them, but on a smaller scale the magnetic field affects the gas, dust and charged particles. Some of the cosmic-ray particles that reach Earth are generated in supernovae explosions within the Milky Way and these are guided to Earth by the galactic magnetic field.

The magnetic field of the Milky Way galaxy yields important information on the structure and behaviour of certain galactic components. The interaction between the magnetic field and the interstellar gas allows us to study low energy plasma physics over galactic dimensions. The interaction between the cosmic-ray particles and the magnetic field constitutes an enormous high energy plasma physics laboratory. These topics of galactic structure not only enhance our general understanding of plasma physics, but also form the springboard from which we can study the magnetic fields of distant galaxies. Since the interstellar magnetic field guides cosmic-ray particles from the rest of the galaxy towards Earth, it is possible to view the entire Milky Way as an extension of the geophysical environment.

Chapter Eight

The Magnetic Fields of Other Galaxies

In recent years it has become clear that magnetic fields exist in many different types of galaxy, and that in certain classes of extragalactic object they play an extremely important role. This chapter reviews briefly what we know of the structure and dynamics of these distant objects, with a particular emphasis on what radio and optical astronomy have together taught us about their magnetic fields. Intergalactic and cosmological magnetic fields are also briefly discussed.

The Realm of the Nebulae

At the turn of the century the arguments still raged over the nature of the nebulae. The application of spectroscopy to astronomical problems had shown that some nebulae were in fact gaseous components of our own Milky Way, but the nature of the spiral nebulae still remained a mystery. The argument was resolved in 1924 by Edwin Hubble, using the 100-inch telescope at Mount Wilson. With this instrument he discovered Cepheid variables in the great Andromeda spiral nebula and in several other spirals. By using the period–luminosity relationship, discussed in the last chapter, he was able to find the distances to these objects.

Using this method as a basis, Hubble employed other methods to find the distances to more distant galaxies for which the Cepheid variable method did not work. One such method made use of supergiant stars which are much brighter intrinsically than Cepheids and so could be seen over larger distances. He assumed that the brightest supergiants in all galaxies have the same absolute brightness, and so observed differences in brightness would be solely due to variations in distance. However, at greater distances even the supergiants could not be used as distance indicators.

At this stage Hubble started using the properties of galaxies themselves. He assumed that all galaxies have the same brightness, thus variations in brightness between galaxies could be used to estimate their distances. Hubble also showed that many—though not all—galaxies tend to occur in clusters: some clusters contain only a few galaxies, while others contain many hundreds.

The Hubble distance scale had to be drastically revised in 1952 when Walter Baade discovered that there are two types of Cepheid variable, and that they have different period–luminosity relationships. All the Cepheids in our own Milky Way galaxy are of Type II, so distances measured using these stars in our own galaxy by Hubble did not need to be revised. However, the Cepheids in other galaxies were of Type I, and this meant that all distances measured using the period–luminosity relation for Type II Cepheids had to be increased by a factor of 5.

For some intermediate distance galaxies it is possible to use novae as distance indicators. A nova is a star which suddenly grows extremely bright and then fades away over a period of a few days. The time taken for a nova to fade to a fraction of its maximum brightness is related to the actual value of this maximum. Novae are occasionally seen in our own galaxy: then it is possible to measure their distances using the techniques described in the previous chapter. From this it is possible to calculate a nova's absolute maximum brightness, and then relate this to the rate at which it fades away. This relationship can be used to find the distances to novae in other galaxies.

Hubble combined his distance measurements with the Doppler measurements of galaxy velocities made by V M Slipher, of the Lowell Observatory, between 1912 and 1924, to deduce Hubble's law on the recession of galaxies. This law states that the speed with which galaxies are moving away from us is directly proportional to their distances from us. This means that distant galaxies are travelling away from our galaxy faster than the nearby ones. Most astronomers now believe that Hubble's law is a direct consequence of the big bang theory for the origin of the universe. According to this theory, all matter in the universe was originally concentrated at very high densities in a small region of space, and this high concentration of matter produced an explosion which sent fragments of matter shooting out in all directions. The galaxies and stars eventually formed out of these fragments of matter. Fragments given the highest speeds by the initial explosion were pushed out furthest, hence one would expect those objects further away from the centre of the explosion to be moving faster than those nearer to the centre. An important consequence of this type of expansion is that from the point of view of any fragment or object in the universe, every other object will be moving away with a speed directly proportional to the distance of each other object from it.

Hubble's law can be used to find the distances to galaxies for which other methods are inadequate. There are either absorption or emission spectral lines in the spectra of most galaxies, and by comparing these with spectral lines generated in a laboratory it is possible, using the Doppler effect, to calculate the speed with which the galaxies are travelling away from our galaxy (figure 8.1). Once this has been done, it is possible to calculate their distances from us using Hubble's law (see figure 8.2). This method is also very useful to radio astronomers since they have no direct way of measuring distances to radio sources.

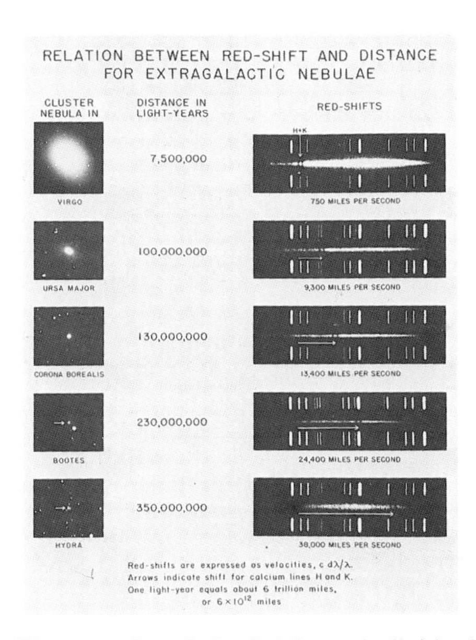

Figure 8.1 Photographs of the brightest individual galaxies in successively more distant clusters of galaxies, shown together with the observed red shift in the light from these galaxies. (Courtesy of the Hale Observatories.)

The system used by optical astronomers to classify galaxies also started with Edwin Hubble. Normal spiral galaxies are similar to our own Milky Way, in that they have two spiral arms surrounding a more compact central nucleus. How tightly the spiral arms are wound up varies from one galaxy to another and leads to further subdivisions within the class (see figure 8.3). Barred spiral galaxies differ from normal spirals in that they have a bar across the centre rather than a nucleus. They also have two spiral arms, but these are not as tightly wound as in normal spirals (see figure 8.4). About 60% of galaxies are normal spirals, and about 20% barred spirals. The rest are a mixture of the other types. Elliptical galaxies are, as their name implies, elliptical in shape. They vary from being almost spherical to highly elliptical, and there are other differences within the class. Irregular galaxies are rather shapeless. The two best known examples of this type are the large and small Magellenic clouds, which can quite easily be seen in the night sky of the southern hemisphere.

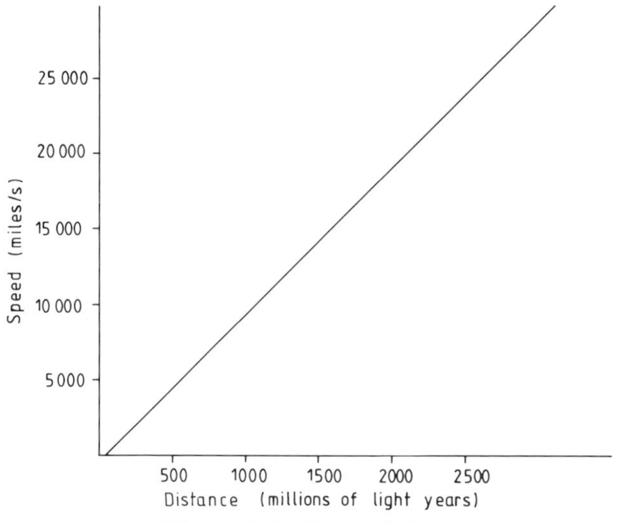

Figure 8.2 Hubble's law.

Quite early in the history of radio, small intense sources of radio radiation were found at fixed points in the sky. By about 1950 the positions of some of these sources were known with enough precision for optical astronomers to find them with giant telescopes. Some of the sources were identified with objects in our Milky Way, while others were identified with irregular galaxies which had unusual properties in their spectra. Radio astronomers soon discovered that the radiation from most of these sources consisted of two or more areas of very strong emission, and that the radiation was highly polarised (see figure 8.5). These sources are now called radio galaxies. The way the intensity of the radiation from these sources varies with the frequency to which the radio telescope is tuned suggests that it arose from the synchrotron process. This is confirmed by the polarisation of the radio galaxies.

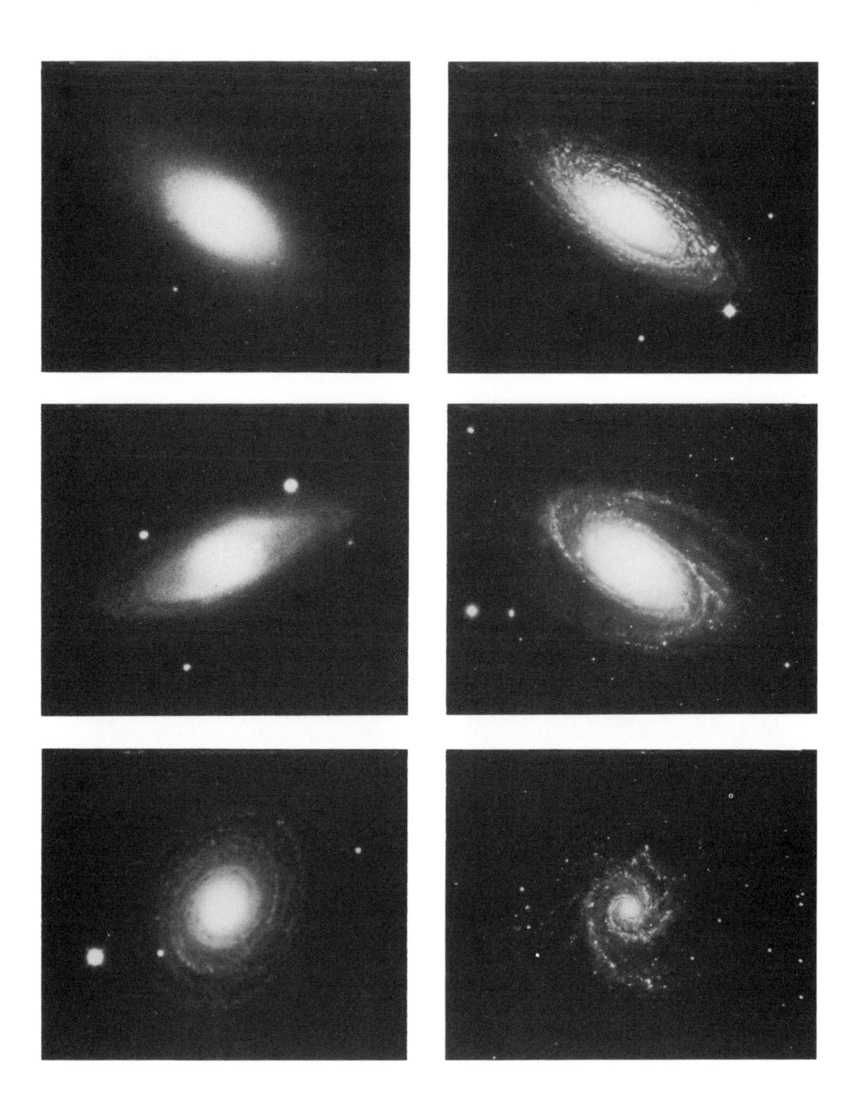

Figure 8.3 Normal spiral galaxies. (Courtesy of the Hale Observatories.)

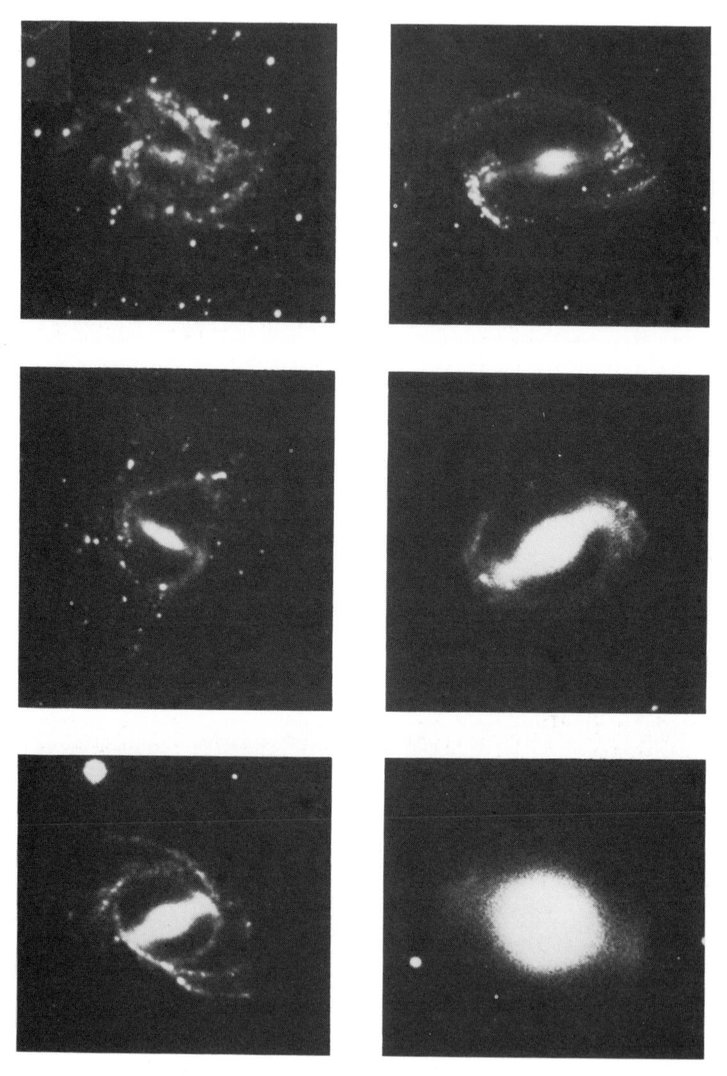

Figure 8.4 Barred spiral galaxies. (Courtesy of the Hale Observatories.)

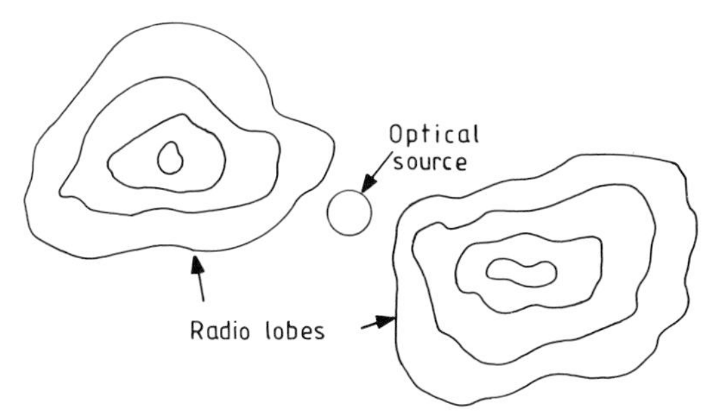

Figure 8.5 The structure of a typical radio galaxy. The curves represent contours of radio intensity.

Quasars

When Hubble's law was applied to the optical spectrum of some of these radio galaxies it showed that they were more distant than most galaxies. Some of these radio sources were in exactly the same part of the sky as optical sources that were so small they looked like stars, but their spectra could only be understood if some of the emission lines were assumed to be shifted very much towards the red end of the spectrum. If Hubble's law was then applied to these redshifts it meant that they must be at very great distances from us, yet they were among the most powerful sources in the radio sky. This suggested that they must be extremely compact and able to produce a great deal of energy in a small volume. Radio astronomers called these sources quasi-stellar radio sources—or quasars.

Their compact nature has been confirmed by two separate sets of observations. The brightnesses of quasars vary on timescales measured in months or weeks. This implies that the quasars are a few light-weeks or light-months across. If this were not the case (i.e. if they were not so compact) light from different parts of the quasar would arrive at the observer at different times and would thus mask the variations. The second set of observations concerns more precise measurements of angular size. The amount of detail that can be 'seen' with a telescope depends on the wavelength being used and the size of the telescope. Because radio wavelengths are longer than optical wavelengths we need telescopes of large dimensions. The sizes of single dish telescopes are limited by engineering considerations. However, this limitation can be overcome by using several separate telescopes, and combining the information collected from a certain object at any particular time by each dish. Such a device is called a radio interferometer. The largest such device available on Earth consists of separate dishes in different continents. An interferometer has been used to confirm that quasars have very small angular sizes. If they are also at great distances from Earth then their actual sizes must also be small.

Over the last few years the intercontinental radio interferometers (or very long baseline interferometers—VLBI for short) have yielded results on quasars that have called into question some basic ideas on the nature of these objects and the fundamental principles of physics. Using VLBIs it has become possible to measure the rate of angular expansion of some quasars. From the angular expansions and the distances deduced from Hubble's law it has been possible to show that some of these sources have an apparent linear speed of expansion which is greater than the speed of light. Since this latter conclusion is not consistent with Einstein's special theory of relativity, some scientists have made renewed efforts to discredit the use of Hubble's law for finding distances to quasars. One of the most ardent of these campaigners is the American astronomer, Halton Arp. He believes that the high redshifts of quasars may be due to other physical causes rather than cosmological expansion. In support of his views he has shown that high speed quasars are often associated with galaxies of much lower redshifts, implying that they are much closer than the distances obtained using Hubble's law. However, most astronomers still accept the cosmological interpretation, and an explanation for expansion speeds greater than the speed of light does now exist. This explanation is based on the changes that take place in measured quantities (e.g. length, time) when matter is travelling close to the speed of light. Let the speed of light be c. If the matter expanding from the centre of quasars is moving in a direction which makes a small angle with the line of sight to the observer, then the transformations associated with Einstein's theory of relativity show that it is possible for parts of the source which are travelling at a substantial fraction of c to give rise to apparent expansion velocities which are greater than c.

Magnetic Fields in Other Galaxies

Optical polarisation measurements have been carried out on several galaxies including the Magellenic clouds, the Andromeda and the Sombrero galaxies. As with the Milky Way, the polarisation of light in these galaxies is believed to be due to the alignment of dust particles in the magnetic fields of these galaxies.

A typical set of results is shown in figure 8.6 for the Sombrero galaxy. This galaxy is seen at a distance of about 40 million miles. It is seen almost edge-on and a prominent feature of the object is the dark lane of dust across its centre. Above and below this dust lane is a lens-shaped luminosity which arises from an oblate distribution of stars. The polarisation map shows fairly regular polarisation close to the lane and parallel to its length. This would imply that the magnetic field of the galaxy is parallel to its central plane. Just above the nucleus the polarisation is at right angles to the lane. Pallister, Scarrott and Bingham, who made the observations, suggest that this is due to reflection of light from the dust grains, rather than the transmission of light through the grains, which is the case for most of the galaxy.

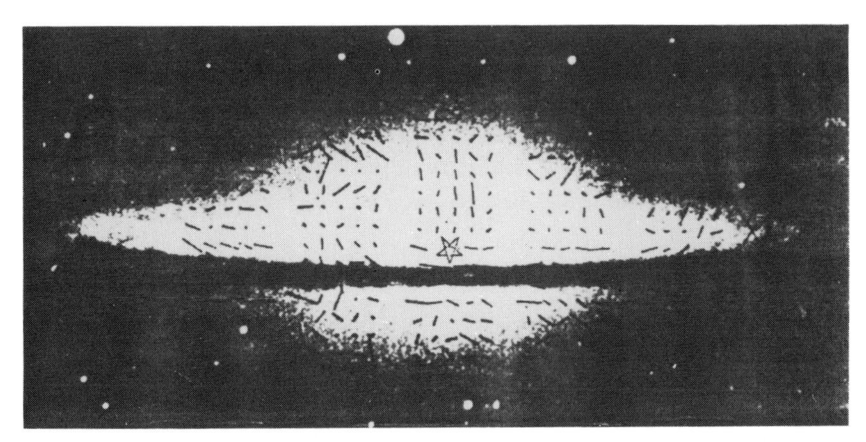

Figure 8.6 Optical polarisation of the Sombrero galaxy. (Reproduced by kind permission of Dr M Scarrott, University of Durham.)

Optical polarisation measurements of spiral galaxies have been supplemented by radio astronomical measurements. By analogy with our own Milky Way it is believed that the radio polarisation results from the synchrotron process at work in these objects. Faraday rotation, the degree and the direction of polarisation (after correction for Faraday rotation) can all be used to discover properties of the field strength and structure. The information obtainable from such studies is illustrated by discussing the Andromeda galaxy.

R Beck, a radio astronomer working at the Max Planck Institute for Radio-astronomy in Germany, made a detailed study of Andromeda. By studying the radio polarisation in this galaxy at 11.1 cm he was able to deduce that this object had a large-scale magnetic field. This field is aligned along the spiral arm segments of neutral hydrogen (figure 8.7). It forms an elliptical ring-like structure in the plane of the galaxy at about 30 000 light years from the centre. His observations showed no large-scale reversals of the magnetic field in the region between 7 and 16 kpc (kiloparsec) (1 kpc = 3260 light years) on a scale equal to the interarm spacing of 2–3 kpc.

A Canadian astronomer, J P Vallée, has used observations on the magnetic fields in spiral galaxies to reach some general conclusions on their structure. He divided these galaxies into two separate classes: one class was composed of galaxies with spiral stellar arms and a spiral magnetic distribution over most of the outer arms; the other class consisted of galaxies with spiral arms and a circular type of magnetic field over large parts of the outer spiral arms. He showed that the first class all had nearby companion galaxies, and suggested that the tidal effects from such a companion could possibly create material arms and align stars, gas and magnetic fields in spiral arms. The second class did not have nearby companions, and consequently had much smaller tidal effects from other galaxies. He also suggested that these smaller tidal effects could only excite density wave arms, with the stars forming a spiral structure, but the gas and magnetic fields forming very nearly circular structures.

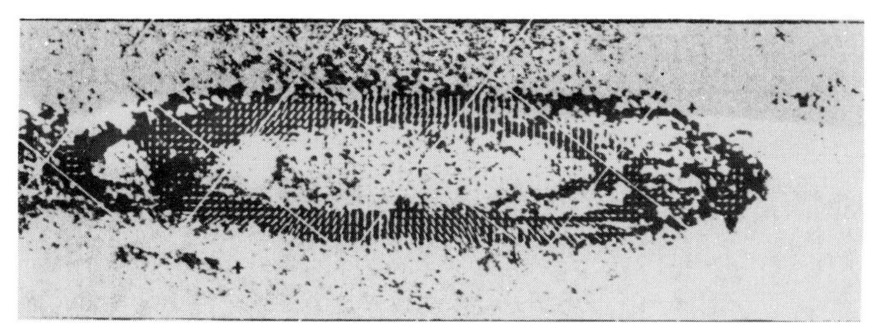

Figure 8.7 Radio polarisation of the Andromeda galaxy. (Reproduced by kind permission of Dr R Beck, Max Planck – Institute for Radio Astronomy, Bonn.)

The polarisation at radio wavelengths of several radio galaxies has also been studied, and a much higher degree of polarisation is shown in these galaxies than in normal spiral galaxies. The strength of the fields in these objects is much higher than in normal galaxies and the fields play a much larger part in the structure and evolution of these radio galaxies than they do in other galaxies. Scientists now believe that the highest energy cosmic-ray particles that reach Earth were probably generated within radio galaxies, and that their magnetic fields play an important role in accelerating these particles to high energies.

Recently a theory has been proposed which explains most of the properties of radio galaxies and quasars. The suggestion is that at the centre of any one of these objects there exists a massive black hole. The black hole will, because of its very high gravitational field, attract matter towards it, but it will also swallow up all the light that is close to its ever-shrinking surface. Suppose that the black hole is rotating and around it is an accretion disc of matter containing a magnetic field (figure 8.8). This magnetic field extends for thousands of light years into space. If the magnetic field were rotating with the black hole it would mean that the outer parts of the field would be moving faster than light.

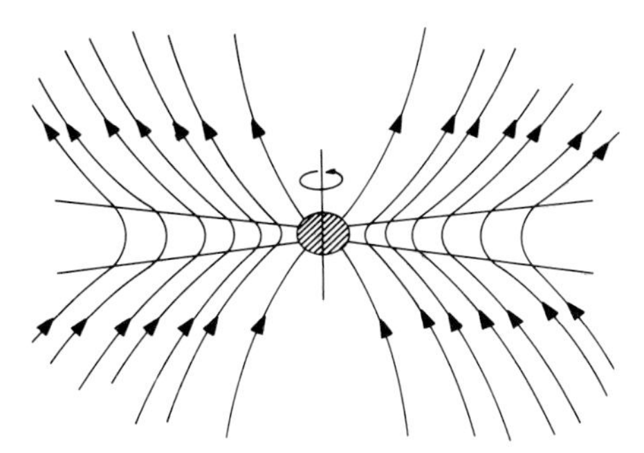

Figure 8.8 Accretion disc around a rotating black hole with a magnetic field.

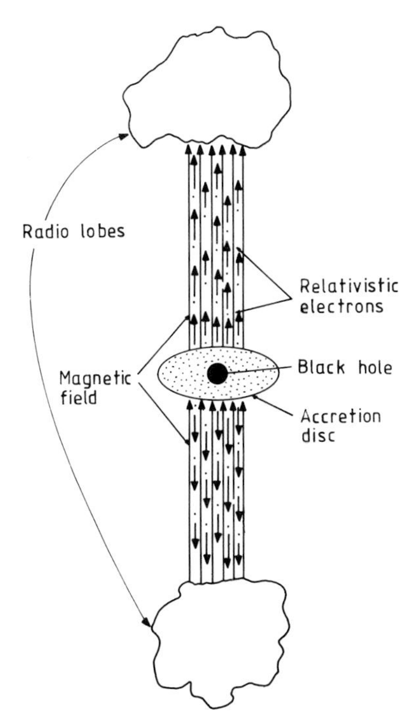

Figure 8.9 Model of a radio galaxy.

Einstein's special theory of relativity tells us that this cannot happen and therefore it has been suggested that the magnetic field 'slips' with respect to the surface of the black hole. The slipping of the field would generate strong electric fields in the neighbourhood of the hole and thus a giant dynamo would be created. According to modern theories in particle physics, electrons and positrons (positively charged electrons) will be created out of the energy of the field. The electric field then propels the electrons along two beams in which they are guided by the magnetic field (figure 8.9). These beams of electrons then give rise to two regions of intense radio emission on either side of a much smaller visible source. This theory is supported by radio and optical studies of a large number of radio galaxies.

Intergalactic and Cosmological Magnetic Fields

At the moment evidence for intergalactic magnetic fields within clusters of galaxies is tenuous. The Milky Way galaxy, the Magellenic clouds, the Andromeda galaxy and about sixteen other galaxies form what is called the Local Group of galaxies. Some years ago two Australian astronomers, Mathewson and Ford, claimed they had found evidence for a magnetic field between the two Magellenic clouds. Subsequent observations by other workers have not confirmed this result, so more convincing evidence is required before it can be stated with any certainty that an intergalactic magnetic field does exist in the Local Group.

There is also no definite evidence for magnetic fields in other clusters of galaxies. In 1968 three Japanese astronomers, Sofue, Fujimoto and Kawabata, claimed they had found evidence for a contribution from an intergalactic medium to the observed rotation measures integrated over elliptical galaxies and quasars. Their suggestion, together with later work by other astronomers, was based on the large-scale asymmetry in the Faraday rotation measures of extragalactic radio sources, and its possible correlation with the redshifts of these sources. They claimed that this could be explained by a weak homogeneous magnetic field stretching over cosmological distances. Vallée, from Canada, has used a much more extensive set of data on Faraday rotation measures to show that this claim is not confirmed by recent results, and that if such a field did exist it must be very weak indeed. However, if such a field does indeed exist, even if it is weak, it must have been stronger in the past, and this being so it could have had consequences for cosmology and the formation of galaxies. It is therefore worth expanding briefly on the possibility of intergalactic magnetic fields and their possible consequences.

The presence of a weak homogeneous field in the universe at the moment implies—because of cosmological expansion—a much stronger homogeneous field in the early universe. Such a homogeneous field would have allowed expansion to take place faster in the direction of the field than at right angles to the field lines. This non-uniform expansion would have observable consequences in that the remnant heat radiation from the early universe would be different in different directions. The high degree of uniformity observed for this background radiation places further limits on the strength of magnetic fields that may have existed in the early universe.

In Chapter Six we saw that the presence of the magnetic field in the interstellar medium can influence the growth of instabilities in the gas clouds out of which stars are formed. In much the same way a magnetic field in the early universe could influence the development of instabilities which lead to the formation of galaxies. Once galaxies are formed the magnetic field could affect the total angular momentum and its distribution within any galaxy. Detailed investigation of these possible effects requires more information on the present strength and uniformity of the field.

Since magnetic fields are not intrinsic properties of space or matter, the existence of a primordial magnetic field does raise questions concerning its origin. Harrison proposed a mechanism by means of which a magnetic field could be generated in an expanding plasma dominated by radiation. These are the kind of conditions that prevailed in the early universe, and can best be described by looking at what happens to a uniformly rotating, expanding whirlpool of matter and radiation, called a vortex. While the vortex is expanding the mass of matter, total density of radiation, and angular momentum of both are all separately conserved. In the plasma, protons collide with the much heavier neutral atoms and as a result the two components of the plasma move with the same angular speed. The protons do not, however, interact with the radiation to the same extent as the electrons, and consequently the radiation is more effective in braking the rotation of the electrons. This means that the protons and electrons move at different speeds, and this gives rise to an electric current which will generate a magnetic field. The application of these principles to the early universe shows that it is possible for a primordial magnetic field to have been generated.

Summary

In this chapter we have seen that magnetic fields exist on scales of galactic dimensions outside our own Milky Way galaxy. In galaxies like our own the stars are kept together by the force of gravity between them, but on a smaller scale the magnetic field affects the gas, dust and charged particles. In radio galaxies the presence of strong magnetic fields plays a strong role in the generation of new particles and also in the large-scale dynamics of these objects. A magnetic field in the early universe may have played a part in its evolution, and in the formation of galaxies. However, the existence of such a field has not yet been convincingly demonstrated.

Technical Appendix

Magnetic Field Strengths in Celestial Objects, and the Units Used to Measure Them

Most measurements in physics are today made in a system of units known as SI units (Système International d'Unités). Just as gravitational field strength is defined as force per unit mass, and electric field strength as force per unit charge, so magnetic field strength (normally called magnetic force density or magnetic induction) is defined as the force per unit current length. In other words it is the force acting per unit length on a conductor which carries unit current and it is at right angles to the direction of the magnetic field. In the SI system, when the force is 1 newton, the current 1 ampere, and the length of the conductor 1 metre, then the strength of the field is 1 tesla. The magnetic field strengths in celestial objects are given in the table overleaf.

Magnetic field strengths in celestial objects.

Celestial object	Field strength (tesla)	Method of measurement
Earth	10^{-4}	Magnetometers on spacecraft
Sun		
general field	10^{-4}	Zeeman effect
sunspots	10^{-1}	Zeeman effect
corona	10^{-5}	Zeeman effect
Solar system		
Mercury	3×10^{-7}	Magnetometers on spacecraft
Jupiter	4×10^{-4}	Magnetometers on spacecraft
Saturn	2×10^{-5}	Magnetometers on spacecraft
Stars		
magnetic stars	1	Zeeman effect
some white dwarfs	10^{2}–10^{4}	Zeeman and synchrotron energy considerations
pulsars	10^{8}	Zeeman and synchrotron energy considerations
x-ray sources near neutron stars	10^{6}–10^{9}	Synchrotron radiation
Galaxy (Milky Way)	10^{-10}	Zeeman, synchrotron, Faraday effect and interstellar polarisation
Radio galaxies and quasars	10^{-2}	Polarisation and Faraday effect

Further Reading

Chapter One Introduction—Why Astronomy?

Man and the Stars R Hanbury-Brown (Oxford: Oxford University Press, 1978)

Chapter Two The Forces of Nature

The Key to the Universe N Calder (London: BBC Publications, 1977)
The Cosmic Onion F Close (London: Heinemann, 1983)
The Forces of Nature P C W Davies (Cambridge: Cambridge University Press, 1979)
The Lighter Side of Gravity J Narliker (San Francisco: Freeman, 1982)
To Acknowledge the Wonder E Squires (Bristol: Adam Hilger, 1985)

Chapter Three The Magnetic Field of the Earth

The Earth's Magnetic Field Unit 23, Science Foundation Course (Open University Press, 1971)
The New Solar System 2nd edn ed J K Beatty, B O'Leary and A Chaikin (Cambridge: Cambridge University Press, 1982)
The Northern Light A Brekke and A Egeland (Berlin: Springer, 1983)

Chapter Four Solar and Interplanetary Magnetic Fields

The Cambridge Encyclopaedia of Astronomy (London: Jonathan Cape, 1977)
Secrets of the Sun R Giovanelli (Cambridge: Cambridge University Press, 1984)
The Sun I Nicolson (London: Mitchell Beazley, 1982)

Chapter Five Magnetic Fields in the Solar System

Jupiter G Hunt and P Moore (London: Mitchell Beazley, 1981)
Introduction to Comets J C Brandt and R D Chapman (Cambridge: Cambridge University Press, 1983)
The Solar System B W Jones (Oxford: Pergamon, 1984)

Chapter Six Magnetic Fields in Stars and Pulsars

Frozen Stars G Greenstein (London: MacDonald, 1983)
Black Holes and Warped Spacetime W J Kaufman (San Francisco: Freeman, 1979)
The Crab Nebula S Mitton (London: Faber and Faber, 1979)

Chapter Seven Magnetic Fields of the Milky Way

The Milky Way B J Bok and P F Bok (Cambridge, MA: Harvard University Press, 1981)

Chapter Eight The Magnetic Fields of Other Galaxies

Exploring the Galaxies S Mitton (London: Faber and Faber, 1976)
Galaxies and Quasars W J Kaufman (San Francisco: Freeman, 1979)

INDEX